Programming Arduino™
Getting Started with Sketches

Simon Monk

New York   Chicago   San Francisco
Lisbon   London   Madrid   Mexico City
Milan   New Delhi   San Juan
Seoul   Singapore   Sydney   Toronto

The *McGraw·Hill* Companies

Library of Congress Cataloging-in-Publication Data

Monk, Simon.
    Programming Arduino : getting started with sketches / Simon Monk.
        p.   cm.
    ISBN 978-0-07-178422-1 (pbk.)
    1. Arduino (Microcontroller)   2. Programmable controllers.   I. Title.
    TJ223.P76M66   2012
    005.13'3—dc23

                                                        2011042537

McGraw-Hill books are available at special quantity discounts to use as premiums
and sales promotions, or for use in corporate training programs. To contact a rep-
resentative, please e-mail us at bulksales@mcgraw-hill.com.

Programming Arduino™: Getting Started with Sketches

Copyright © 2012 by The McGraw-Hill Companies. All rights reserved. Printed in
the United States of America. Except as permitted under the Copyright Act of
1976, no part of this publication may be reproduced or distributed in any form or
by any means, or stored in a database or retrieval system, without the prior written permission of publisher, with the exception that the program listings may be
entered, stored, and executed in a computer system, but they may not be reproduced for publication.

All trademarks or copyrights mentioned herein are the possession of their respec-
tive owners and McGraw-Hill makes no claim of ownership by the mention of
products that contain these marks.

"Arduino" is a trademark of the Arduino team.

4 5 6 7 8 9 0   DOC DOC   1 5 4 3 2

ISBN   978-0-07-178422-1

MHID   0-07-178422-5

Sponsoring Editor	**Copy Editor**	**Composition**
Roger Stewart	Andy Saff	Cenveo Publisher Services
Editorial Supervisor	**Proofreader**	
Jody McKenzie	Carol Shields	**Illustration**
Project Manager	**Indexer**	Cenveo Publisher Services
Sapna Rastogi, Cenveo Publisher Services	Jack Lewis	**Art Director, Cover**
	Production Supervisor	Jeff Weeks
Acquisitions Coordinator	George Anderson	
Joya Anthony		

Information has been obtained by McGraw-Hill from sources believed to be reliable. However, because of
the possibility of human or mechanical error by our sources, McGraw-Hill, or others, McGraw-Hill does
not guarantee the accuracy, adequacy, or completeness of any information and is not responsible for any
errors or omissions or the results obtained from the use of such information.

To my boys, Stephen and Matthew,
from a very proud Dad.

About the Author

Simon Monk has a bachelor's degree in cybernetics and computer science and a doctorate in software engineering. He has been an active electronics hobbyist since his school days and is an occasional author in hobby electronics magazines. He is also author of *30 Arduino Projects for the Evil Genius* and *15 Dangerously Mad Projects for the Evil Genius.*

CONTENTS

ACKNOWLEDGMENTS

I thank Linda for giving me the time, space, and support to write this book and for putting up with the various messes my projects create around the house.

I also thank Stephen and Matthew Monk for taking an interest in what their Dad is up to and their general assistance with project work.

Finally, I would like to thank Roger Stewart, Sapna Rastogi, and everyone involved in the production of this book. It's a pleasure to work with such a great team.

INTRODUCTION

Arduino interface boards provide a low-cost, easy-to-use technology to create microcontroller-based projects. With a little electronics, you can make your Arduino do all sorts of things, from controlling lights in an art installation to managing the power on a solar energy system.

There are many project-based books that show you how to connect things to your Arduino, including *30 Arduino Projects for the Evil Genius* by this author. However, the focus of this book is on programming the Arduino.

This book will explain how to make programming the Arduino simple and enjoyable, avoiding the difficulties of uncooperative code that so often afflict a project. You will be taken through the process of programming the Arduino step by step, starting with the basics of the C programming language that Arduinos use.

So, What Is Arduino?

Arduino is a small microcontroller board with a universal serial bus (USB) plug to connect to your computer and a number of connection sockets that can be wired to external electronics such as motors, relays, light sensors, laser diodes, loudspeakers, microphones, and more. They can either be powered through the USB connection from the computer, from a 9V battery, or from a power supply. They can be controlled from the computer or programmed by the computer and then disconnected and allowed to work independently.

The board design is open source. This means that anyone is allowed to make Arduino-compatible boards. This competition has lead to low costs for the boards.

The basic boards are supplemented by accessory shield boards that can be plugged on top of the Arduino board. In this book, we will use two shields—an LCD display shield and an Ethernet shield—that will allow us to turn our Arduino into a tiny web server.

The software for programming your Arduino is easy to use and also freely available for Windows, Mac, and LINUX computers.

What Will I Need?

This is a book intended for beginners, but it is also intended to be useful to those who have used Arduino for a while and want to learn more about programming the Arduino or gain a better understanding of the fundamentals.

You do not need to have any programming experience or a technical background, and the book's exercises do not require any soldering. All you need is the desire to make something.

If you want to make the most of the book and try out some of the experiments, then it is useful to have the following on hand:

- A few lengths of solid core wire
- A cheap digital multimeter

Both are readily available for a few dollars from a hobby electronics shop such as Radio Shack. You will of course also need an Arduino Uno board.

If you want to go a step further and experiment with Ethernet and the liquid crystal display (LCD) shield, then you will need to buy shields that are available from online stores. See Chapters 9 and 10 for details.

Using this Book

This book is structured to get you started in a really simple way and gradually build on what you have learned. You may, however, find yourself skipping or skimming some of the early chapters as you find the right level to enter the book.

The book is organized into the following chapters:

- **Chapter 1: This Is Arduino**   An introduction to the Arduino hardware, this chapter describes what it is capable of, and the various types of, Arduino boards that are available.

- **Chapter 2: Getting Started**   Here you conduct your first experiments with your Arduino board: installing the software, powering it up, and uploading your first sketch.

- **Chapter 3: C Language Basics**   This chapter covers the basics of the C language; for complete programming beginners, the chapters also serves as an introduction to programming in general.

- **Chapter 4: Functions**   This chapter explains the key concept of using and writing functions in Arduino sketches. These sketches are demonstrated throughout with runnable code examples.

- **Chapter 5: Arrays and Strings**   Here you learn how to make and use data structures that are more advanced than simple integer variables. A Morse code example project is slowly developed to illustrate the concepts being explained.

- **Chapter 6: Input and Output**   You learn how to use the digital and analog inputs and outputs on the Arduino in your programs. A multimeter will be useful to show you what is happening on the Arduino's input/output connections.

- **Chapter 7: The Standard Arduino Library**   This chapter explains how to make use of the standard Arduino functions that come in the Arduino's standard library.

- **Chapter 8: Data Storage**   Here you learn how to write sketches that can save data in electrically erasable read-only memory (EEPROM) and make use of the Arduino's built-in flash memory.

- **Chapter 9: LCD Displays**   In this chapter, you program with the LCD Shield library to make a simple USB message board example.

- **Chapter 10: Arduino Ethernet Programming**   You learn how to make the Arduino behave like a web server as you get a little background on HyperText Markup Language (HTML) and the HyperText Transfer Protocol (HTTP).

- **Chapter 11: C++ and Libraries**   You go beyond C, looking at adding object-orientation and writing your own Arduino libraries.

Resources

This book is supported by an accompanying website:

www.arduinobook.com

There you will find all the source code used in this book as well as other resources, such as errata.

1

This Is Arduino

Arduino is a microcontroller platform that has captured the imagination of electronics enthusiasts. Its ease of use and open source nature make it a great choice for anyone wanting to build electronic projects.

Ultimately, it allows you to connect electronics through its pins so that it can control things—for instance, turn lights or motors on and off or sense things such as light and temperature. This is why Arduino is sometimes given the description *physical computing*. Because Arduinos can be connected to your computer by a universal serial bus (USB) lead, this also means that you can use the Arduino as an interface board to control those same electronics from your computer.

This chapter is an introduction to the Arduino, including the history and background of the Arduino, as well as an overview of the hardware.

Microcontrollers

The heart of your Arduino is a microcontroller. Pretty much everything else on the board is concerned with providing the board with power and allowing it to communicate with your desktop computer.

A microcontroller really is a little computer on a chip. It has everything and more than the first home computers had. It has a processor, a kilobyte or two of random access memory (RAM) for holding data, a few kilobytes of erasable programmable read-only memory (EPROM) or flash memory for holding your programs and it has input and output pins. These input/output (I/O) pins link the microcontroller to the rest of your electronics.

1

Inputs can read both digital (is the switch on or off?) and analog (what is the voltage at a pin?). This opens up the opportunity of connecting many different types of sensor for light, temperature, sound, and more.

Outputs can also be analog or digital. So, you can set a pin to be on or off (0 volts or 5 volts) and this can turn light-emitting diodes (LEDs) on and off directly, or you can use the output to control higher power devices such as motors. They can also provide an analog output voltage. That is, you can set the output of a pin to some particular voltage, allowing you to control the speed of a motor or the brightness of a light, rather than simply turning it on or off.

The microcontroller on an Arduino board is the 28-pin chip fitted into a socket at the center of the board. This single chip contains the memory processor and all the electronics for the input/output pins. It is manufactured by the company Atmel, which is one of the major microcontroller manufacturers. Each of the microcontroller manufacturers actually produces dozens of different microcontrollers grouped into different families. The microcontrollers are not all created for the benefit of electronics hobbyists like us. We are a small part of this vast market. These devices are really intended for embedding into consumer products, including cars, washing machines, DVD players, children's toys, and even air fresheners.

The great thing about the Arduino is that it reduces this bewildering array of choices by standardizing on one microcontroller and sticking with it. (Well, as we see later, this statement is not quite true, but it's close enough.)

This means that when you are embarking on a new project, you do not first need to weigh all the pros and cons of the various flavors of microcontroller.

Development Boards

We have established that the microcontroller is really just a chip. A chip will not just work on its own without some supporting electronics to provide it with a regulated and accurate supply of electricity (microcontrollers are fussy about this) as well as a means of communicating with the computer that is going to program the microcontroller.

This is where development boards come in. An Arduino board is really a microcontroller development board that happens to be an independent open source hardware design. This means that the design files for the printed circuit board (PCB) and the schematic diagrams are all publicly available, and everyone is free to use the designs to make and sell his or her own Arduino boards.

All the microcontroller manufacturers—including Atmel, which makes the ATmega328 microcontroller used in an Arduino board—also provide their own development boards and programming software. Although they are usually fairly inexpensive, these tend to be aimed at professional electronics engineers rather than hobbyists. This means that such boards and software are arguably harder to use and require a greater learning investment before you can get anything useful out of them.

A Tour of an Arduino Board

Figure 1-1 shows an Arduino board. Let's take a quick tour of the various components on the board.

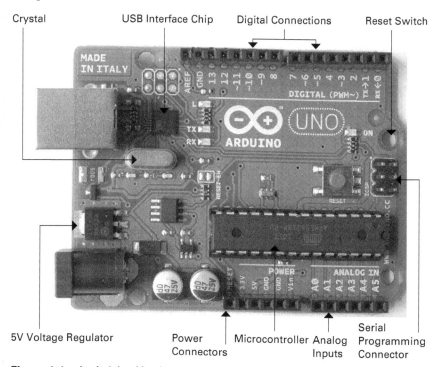

Figure 1-1 *An Arduino Uno board*

Power Supply

Referring to Figure 1-1, directly below the USB connector is the 5-volt (5V) voltage regulator. This regulates whatever voltage (between 7V and 12V) is supplied from the power socket into a constant 5V.

The 5V voltage regulator chip is actually quite big for a surface mount component. This is so that it can dissipate the heat required to regulate the voltage at a reasonably high current. This is useful when driving external electronics.

Power Connections

Next let us look at the connectors at the bottom of Figure 1-1. You can read the connection names next to the connectors. The first is Reset. This does the same thing as the Reset button on the Arduino. Rather like rebooting a PC, using the Reset connector resets the microcontroller so that it begins its program from the start. To reset the microcontroller with the Reset connector, you momentarily set this pin low (connecting it to 0V).

The rest of the pins in this section just provide different voltages (3.5V, 5V, GND, and 9V), as they are labeled. GND, or ground, just means zero volts. It is the reference voltage to which all other voltages on the board are relative.

Analog Inputs

The six pins labeled as Analog In A0 to A5 can be used to measure the voltage connected to them so that the value can be used in a sketch. Note that they measure a voltage and not a current. Only a tiny current will ever flow into them and down to ground because they have a very large internal resistance. That is, the pin having a large internal resistance only allows a tiny current to flow into the pin.

Although these inputs are labeled as analog, and are analog inputs by default, these connections can also be used as digital inputs or outputs.

Digital Connections

We now switch to the top connector and start on the right-hand side in Figure 1-1. Here we find pins labeled Digital 0 to 13. These can be used as either inputs or outputs. When used as outputs, they behave rather like the power supply voltages discussed earlier in this section, except that these are all 5V and can be turned on or off from your sketch. So, if you turn them on from your sketch they will be at 5V, and if you turn them off they will be at 0V. As with the power supply connectors, you must be careful not to exceed their maximum current capabilities. The first two of these connections (0 and 1) are also labeled RX and TX, for receive and transmit. These connections are reserved for use in communication and are indirectly the receive and transmit connections for your USB link to your computer.

These digital connections can supply 40 mA (milliamps) at 5V. That is more than enough to light a standard LED, but not enough to drive an electric motor directly.

Microcontroller

Continuing our tour of the Arduino board, the microcontroller chip itself is the black rectangular device with 28 pins. This is fitted into a dual in-line (DIL) socket so that it can be easily replaced. The 28-pin microcontroller chip used on the Arduino Uno board is the ATmega328. Figure 1-2 is a block diagram showing the main features of this device.

The heart—or, perhaps more appropriately, the brain—of the device is the central processing unit (CPU). It controls everything that goes on within the device. It fetches program instructions stored in the flash memory and executes them. This might involve fetching data from working memory (RAM), changing it, and then putting it back. Or, it may mean changing one of the digital outputs from 0V to 5V.

The EEPROM memory is a little like the flash memory in that it is non-volatile. That is, you can turn the device off and on and it will not have forgotten what is in the EEPROM. Whereas the flash memory is intended

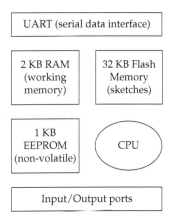

Figure 1-2 *ATmega328 block diagram*

for storing program instructions (from sketches), the EEPROM is used to store data that you do not want to lose in the event of a reset or the power being turned off.

Other Components

Above the microcontroller is a small, silver, rectangular component. This is a quartz crystal oscillator. It ticks 16 million times a second, and on each of those ticks, the microcontroller can perform one operation—addition, subtraction, or another mathematical operation.

To the right of the crystal is the Reset switch. Clicking on this switch sends a logic pulse to the Reset pin of the microcontroller, causing the microcontroller to start its program afresh and clear its memory. Note that any program stored on the device will be retained, because this is kept in non-volatile flash memory—that is, memory that remembers even when the device is not powered.

To the right of the Reset button is the Serial Programming Connector. It offers another means of programming the Arduino without using the USB port. Because we do have a USB connection and software that makes it convenient to use, we will not avail ourselves of this feature.

In the top-left corner of the board next to the USB socket is the USB interface chip. This chip converts the signal levels used by the USB standard to levels that can be used directly by the Arduino board.

The Origins of Arduino

Arduino was originally developed as an aid for teaching students. It was subsequently (in 2005) developed commercially by Massimo Banzi and David Cuartielles. It has since gone on to become enormously successful with makers, students, and artists for its ease of use and durability.

Another key factor in its success is that all the designs for Arduino are freely available under a Creative Commons license. This has allowed many lower-cost alternatives to the boards to appear. Only the name Arduino is protected, so such clones often have "*dunino" names, such as Boarduino, Seeeduino, and Freeduino. However, the official boards manufactured in Italy still sell extremely well. Many big retailers sell only the official boards, which are nicely packaged and of high quality.

Yet another reason for the success of Arduino is that it is not limited to microcontroller boards. There are a huge number of Arduino-compatible shield boards that plug directly into the top of an Arduino board. Because shields are available for almost every conceivable application, you often can avoid using a soldering iron and instead plug together shields that can be stacked one upon another. The following are just a few of the most popular shields:

- Ethernet, which gives an Arduino web-serving capabilities
- Motor, which drives electric motors
- USB Host, which allows control of USB devices
- Relays, which switches relays from your Arduino

Figure 1-3 shows an Arduino Uno with an Ethernet shield attached.

Figure 1-3 *Arduino Uno with an Ethernet shield*

The Arduino Family

It is useful to have a little background on the various Arduino boards. We will be using the Arduino Uno as our standard device. Indeed, this is by far the most used of the Arduino boards, but the boards are all programmed using the same language and largely have the same connections to the outside world, so you can easily use a different board.

Uno, Duemilanove, and Diecimila

The Arduino Uno is the latest incarnation of the most popular series of Arduino boards. The series includes the Diecimila (Italian for 10,000) and the Duemilanove (Italian for 2009). Figure 1-4 shows an Arduino clone. By now you may have guessed that Arduino is an Italian invention.

These older boards look very similar to the Arduino Uno. They both have the same connectors and a USB socket and are generally compatible with each other.

The most significant difference between the Uno and the earlier boards is that the Uno uses a different USB chip. This does not affect how you use the board, but it does make installation of the Arduino software easier and allows higher speeds of communication with the computer.

The Uno can also supply more current on its 3.3V supply and always comes equipped with the ATmega328. The earlier boards will have either an ATmega328 or ATmega168. The ATmega328 has more memory, but unless you are creating a large sketch, this will make no difference.

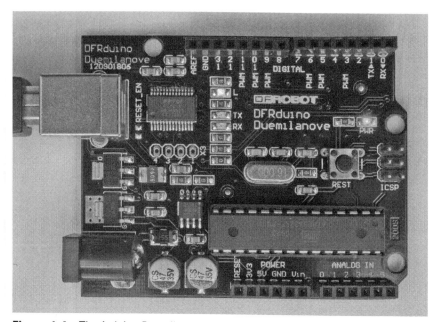

Figure 1-4   *The Arduino Duemilanove*

Mega

The Arduino Mega (Figure 1-5) is the muscle car of Arduino boards. It boasts a huge collection of input output ports, but cleverly adds these as extra connectors at one end of the board, allowing it to remain pin-compatible with the Arduino Uno and all the shields available for Arduino.

It uses a processor with more input output pins, the ATmega1280, which is a surface mount chip that is fixed permanently to the board. So, unlike with the Uno and similar boards, you cannot replace the processor if you accidentally damage it.

The extra connectors are arranged at the end of the board. Extra features provided by the Mega include the following:

- 54 input/output pins
- 128KB of flash memory for storing sketches and fixed data (compared to the Uno's 32KB)
- 8KB of RAM and 4KB of EEPROM

Nano

The Arduino Nano (Figure 1-6) is a very useful device for use with a solderless breadboard. If you fit pins to it, it can just plug into the breadboard as if it were a chip.

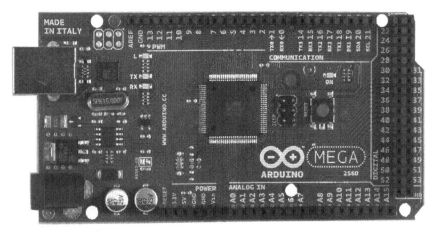

Figure 1-5 *An Arduino Mega board*

Figure 1-6   *Arduino Nano*

The downside of the Nano is that because it is so much smaller than an Uno, it cannot accept Uno-sized shields.

Bluetooth

The Arduino Bluetooth (Figure 1-7) is an interesting device as it includes Bluetooth hardware in place of the USB connector. This allows the device to even be programmed wirelessly.

The Arduino Bluetooth is not a cheap board, and it is often cheaper to attach a third-party Bluetooth module to a regular Arduino Uno.

Lilypad

The Lilypad (Figure 1-8) is a tiny, thin Arduino board that can be stitched into clothing for applications that have become known as wearable computing.

The Lilypad does not have a USB connection, and you must use a separate adaptor to program it. It has an exceptionally beautiful design.

Figure 1-7 *Arduino Bluetooth*

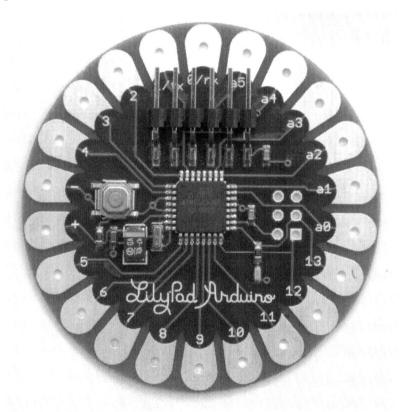

Figure 1-8 *Arduino Lilypad*

Other "Official" Boards

The previously described Arduino boards are the most useful and popular ones. However, the range of Arduino boards constantly changes, so for a complete and up-to-date picture of the Arduino family, see the official Arduino website list at www.arduino.cc/en/Main/Hardware.

Arduino Clones and Variants

Unofficial boards fall into two categories. Some just take the standard open source hardware designs of Arduino and build a cheaper one. Some names you can search for boards of this nature include the following:

- Roboduino
- Freeduino
- Seeeduino (yes, with three *e*'s)

More interestingly, some Arduino-compatible designs are intended to extend or improve the Arduino in some way. New variants are appearing all the time, and far too many exist to mention them all. However, the following are some of the more interesting and popular variants:

- Chipkit, a high-speed variant based on a PIC processor, but which is fairly compatible with Arduino
- Femtoduino, a very small Arduino
- Ruggeduino, with is an Arduino board with built-in I/O protection
- Teensy, a low-cost nano-type device

Conclusion

Now that you have explored the Arduino hardware a little, it's time to set up your Arduino software.

2

Getting Started

Having introduced the Arduino, and learnt a little about what it is that we are programming, it is time to learn how install the software that we will need on our computer and to start working on some code.

Powering Up

When you buy an Arduino board, it is usually preinstalled with a sample Blink program that will make the little built-in light-emitting diode (LED) flash.

The LED marked L is wired up to one of the digital input output sockets on the board. It is connected to digital pin 13. This limits pin 13 to being the one used as an output. However, the LED uses only a small amount of current, so you can still connect other things to that connector.

All you need to do to get your Arduino up and running is supply it with some power. The easiest way to do this is to plug in it into the USB port on your computer. You will need a type-A-to-type-B USB lead. This is the same type of lead that is normally used to connect a computer to a printer.

If everything is working OK, the LED should blink. New Arduino boards come with this Blink sketch already installed so that you can verify that the board works.

Installing the Software

To be able to install new sketches onto your Arduino board, you need to do more than supply power to it over the USB. You need to install the Arduino software (Figure 2-1).

Full and comprehensive instructions for installing this software on Windows, Linux, and Mac computers can be found at the Arduino website (www.arduino.cc).

Once you have successfully installed the Arduino software and, depending on your platform, USB drivers, you should now be able to upload a program to the Arduino board.

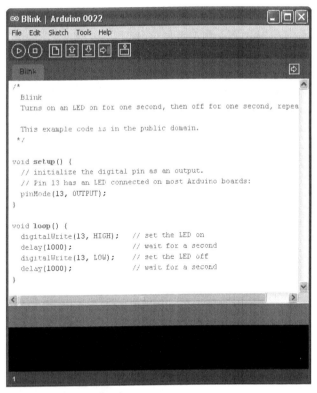

Figure 2-1   *The Arduino application*

Uploading Your First Sketch

The blinking LED is the Arduino equivalent to the "Hello World" program used in other languages as the traditional first program to run when learning a new language. Let's test out the environment by installing this program on your Arduino board and then modifying it.

When you start the Arduino application on your computer, it opens with an empty sketch. Fortunately, the application ships with a wide range of useful examples. So from the File menu, open the Blink example as shown in Figure 2-2.

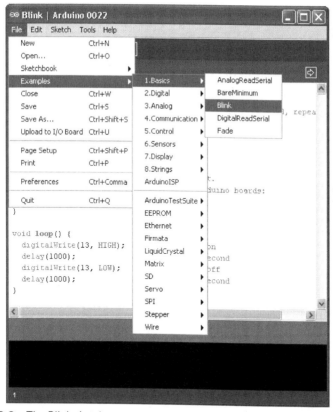

Figure 2-2 *The Blink sketch*

You now need to transfer or upload that sketch to your Arduino board. So plug your Arduino board into your computer using the USB lead. You should see the green "On" LED on the Arduino light up. The Arduino board will probably already be flashing, as the boards are generally shipped with the Blink sketch already installed. But let's install it again and then modify it.

When you plug the board in, if you are using a Mac, you may get the message, "A new network interface has been detected." Just click Cancel; your Mac is confused and thinks that the Uno is a USB modem.

Before you can upload a sketch, you must tell the Arduino application what type of board you are using and which serial port you are connected to. Figures 2-3 and 2-4 show how you do this from the Tools menu.

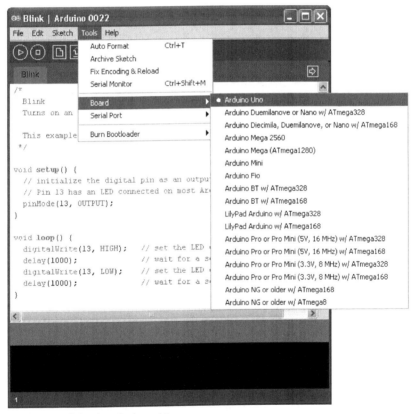

Figure 2-3 *Selecting the board type*

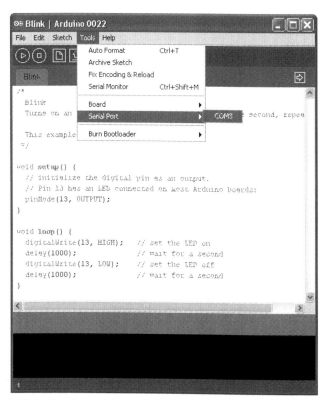

Figure 2-4 *Selecting the serial port (in Windows)*

On a Windows machine, the serial port is always COM3. On Macs and Linux machines, you will see a much longer list of serial devices (see Figure 2-5). The device will normally be the top selection in the list, with a name similar to /dev/tty.usbmodem621.

Now click on the Upload icon in the toolbar. This is shown highlighted in Figure 2-6.

After you click the button, there is a short pause while the sketch is compiled and then the transfer begins. If it is working, then there will be some furious blinking of LEDs as the sketch is transferred, after which

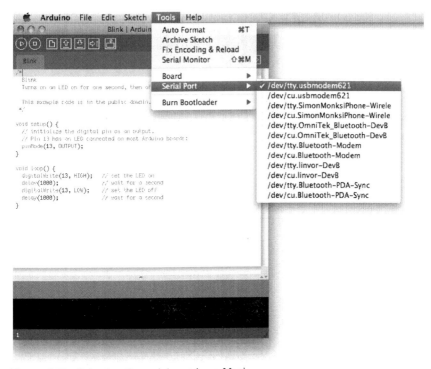

Figure 2-5 *Selecting the serial port (on a Mac)*

you should see the message "Done Uploading" at the bottom of the Arduino application window and a further message similar to "Binary sketch size: 1018 bytes (of a 14336 byte maximum)."

Once uploaded, the board automatically starts running the sketch and you will see the LED start to blink.

If this did not work, then check your serial and board type settings.

Now let's modify the sketch to make the LED blink faster. To do this, let's alter the two places in the sketch where there is a delay for 1,000 milliseconds

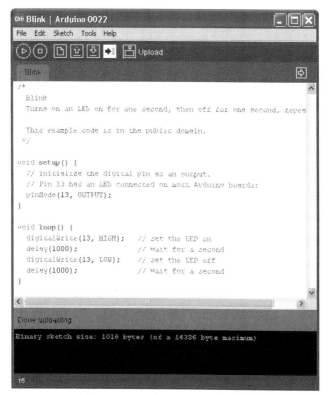

```
Blink
Turns on an LED on for one second, then off for one second, repea

This example code is in the public domain.
*/

void setup() {
  // initialize the digital pin as an output.
  // Pin 13 has an LED connected on most Arduino boards:
  pinMode(13, OUTPUT);
}

void loop() {
  digitalWrite(13, HIGH);   // set the LED on
  delay(1000);              // wait for a second
  digitalWrite(13, LOW);    // set the LED off
  delay(1000);              // wait for a second
}
```

Figure 2-6 *Uploading the sketch*

so that the delay is 500 milliseconds. Figure 2-7 shows the modified sketch with the changes highlighted.

Click on the Upload button again. Then, once the sketch has uploaded, you should see your LED start to blink twice as fast as it did before.

Congratulations, you are now ready to start programming your Arduino. First, though, let's take a mini-tour of the Arduino application.

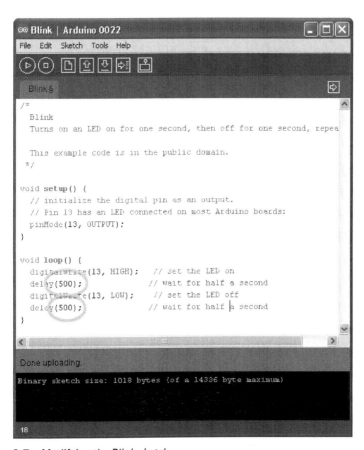

Figure 2-7 *Modifying the Blink sketch*

The Arduino Application

Sketches in Arduino are like documents in a word processor. You can open them and copy parts from one to another. So you see options to Open, Save, and Save As in the File menu. You will not normally use Open because the Arduino application has the concept of a Sketchbook where all your sketches are kept carefully organized into folders. You gain access to the Sketchbook from the File menu. As you have just installed the

Arduino application for the first time, your Sketchbook will be empty until you create some sketches.

As you have seen, the Arduino application comes with a selection of example sketches that can be very useful. Having modified the Blink example sketch, if you try and save it, you get a message that says, "Some files are marked read-only so you will need to save this sketch in a different location."

Try this now. Accept the default location, but change the filename to MyBlink, as shown in Figure 2-8.

Now if you go to the File menu and then click on Sketches, you will see MyBlink as one of the sketches listed. If you look at your computer's file system, on a PC, you will find that the sketch has been written into My Documents\Arduino, and on Mac or Linux, they are in Documents/Arduino.

All of the sketches used in this book can be downloaded as a zip file (Programming_Arduino.zip) from www.arduinobook.com. I suggest that

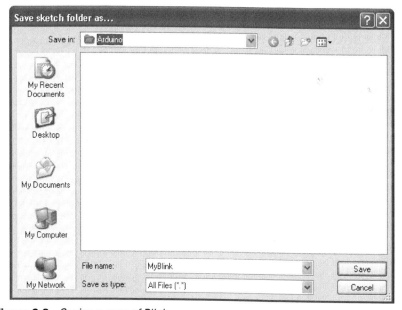

Figure 2-8 *Saving a copy of Blink*

now is the time to download this file and unzip it into the Arduino folder that contains the sketches. In other words, when you have unzipped the folder, there should be two folders in your Arduino folder: one for the newly saved MyBlink and one called Programming Arduino (see Figure 2-9). The Programming Arduino folder will contain all the sketches, numbered according to chapter, so that sketch 03-01, for example, is sketch 1 of Chapter 3.

These sketches will not appear in your Sketchbook menu until you quit the Arduino application and restart it. Do so now. Then your Sketchbook menu should look similar to that shown in Figure 2-10.

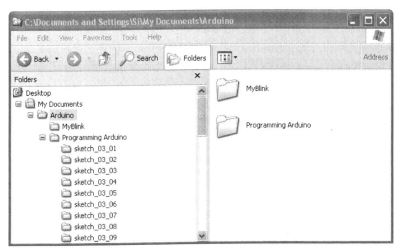

Figure 2-9 *Installing the sketches from the book*

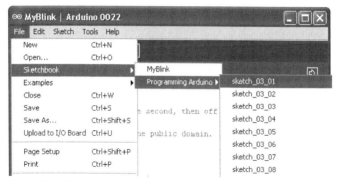

Figure 2-10   *Sketchbook with the book's sketches installed*

Conclusion

Your environment is all set up and ready to go.

In the next chapter, we will look at some of the basic principles of the C language that the Arduino uses and start writing some code.

3

C Language Basics

The programming language used to program Arduinos is a language called C. In this chapter, you get to understand the basics of the C language. You will use what you learn here in every sketch you develop as an Arduino programmer. To get the most out of Arduino, you need to understand these fundamentals.

Programming

It is not uncommon for people to speak more than one language. In fact, the more you learn, the easier it seems to learn spoken languages as you start to find common patterns of grammar and vocabulary. The same is true of programming languages. So, if you have used any other programming language, you will quickly pick up C.

The good news is that the vocabulary of a programming language is far smaller than that of a spoken language, and because you write it rather than say it, the dictionary can always be at hand whenever you need to look things up. Also, the grammar and syntax of a programming language are extremely regular, and once you come to grips with a few simple concepts, learning more quickly becomes second nature.

It is best to think of a program—or a sketch, as programs are called in Arduino—as a list of instructions to be carried out in the order that they are written down. For example, suppose you were to write the following:

```
digitalWrite(13, HIGH);
delay(500);
digitalWrite(13, LOW);
```

These three lines would each do something. The first line would set the output of pin 13 to HIGH. This is the pin with an LED built in to the Arduino board, so at this point the LED would light. The second line would simply wait for 500 milliseconds (half a second) and then the third line would turn the LED back off again. So these three lines would achieve the goal of making the LED blink once.

You have already seen a bewildering array of punctuation used in strange ways and words that don't have spaces between them. A frustration of many new programmers is, "I know what I want to do, I just don't know what I need to write!" Fear not, all will be explained.

First of all, let's deal with the punctuation and the way the words are formed. These are both part of what is termed the syntax of the language. Most languages require you to be extremely precise about syntax, and one of the main rules is that names for things have to be a single word. That is, they cannot include spaces. So, **digitalWrite** is the name for something. It's the name of a built-in function (you'll learn more about functions later) that will do the job of setting an output pin on the Arduino board. Not only do you have to avoid spaces in names, but also names are case sensitive. So you must write **digitalWrite**, not **DigitalWrite** or **Digitalwrite**.

The function **digitalWrite** needs to know which pin to set and whether to set that pin HIGH or LOW. These two pieces of information are called *arguments*, which are said to be *passed* to a function when it is *called*. The parameters for a function must be enclosed in parentheses and separated by commas.

The convention is to place the opening parenthesis immediately after the last letter of the function's name and to put a space after the comma before the next parameter. However, you can sprinkle space characters within the parentheses if you want.

If the function only has one argument, then there is no need for a comma. Notice how each line ends with a semicolon. It would be more logical if they were periods, because the semicolon marks the end of one command, a bit like the end of a sentence.

In the next section, you will find out a bit more about what happens when you press the Upload button on the Arduino integrated development environment (IDE). Then you will be able to start trying out a few examples.

What Is a Programming Language?

It is perhaps a little surprising that we can get to Chapter 3 in a book about programming without defining exactly what a programming language is. We can recognize an Arduino sketch and probably have a rough idea of what it is trying to do, but we need to look a bit deeper into how some programming language code goes from being words on a page to something that does something real, like turn an LED on and off.

Figure 3-1 summarizes the process involved from typing code into the Arduino IDE to running the sketch on the board.

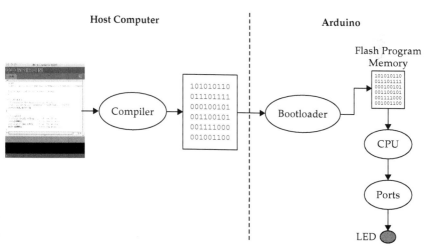

Figure 3-1 *From code to board*

When you press the Upload button on your Arduino IDE, it launches a chain of events that results in your sketch being installed on the Arduino and being run. This is not as straightforward as simply taking the text that you typed into the editor and moving it to the Arduino board.

The first step is to do something called *compilation*. This takes the code you have written and translates it into machine code—the binary language that the Arduino understands. If you click the triangular Verify button on the Arduino IDE, this actually attempts to compile the C that you have written without trying to send the code to the Arduino IDE. A side-effect of compiling the code is that it is checked to make sure that it conforms to the rules of the C language.

If you type **Ciao Bella** into your Arduino IDE and click on the Play button, the results will be as shown in Figure 3-2.

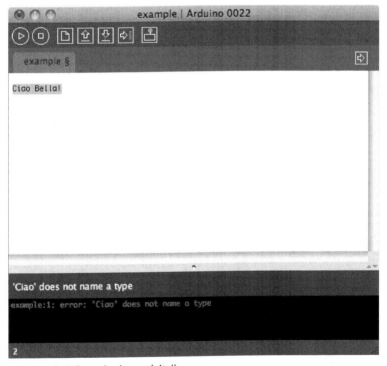

Figure 3-2 *Arduinos don't speak Italian*

The Arduino has tried to compile the words "Ciao Bella," and despite its Italian heritage, it has no idea what you are talking about. This text is not C. So, the result is that at the bottom of the screen we have that cryptic message "error: Ciao does not name a type." What this actually means is that there is a lot wrong with what you have written.

Let's try another example. This time we will try compiling a sketch with no code at all in it (see Figure 3-3).

This time, the compiler is telling you that your sketch does not have **setup** or **loop** functions. As you know from the Blink example that you ran in Chapter 2, you have to have some "boilerplate" code, as it is called, before you can add your own code into a sketch. In Arduino programming the "boilerplate" code takes the form of the "setup" and "loop" functions that must always be present in a sketch.

Figure 3-3 *No* setup *or* loop

You will learn much more about functions later in the book, but for now, let's accept that you need this boilerplate code and just adapt your sketch so it will compile (see Figure 3-4).

The Arduino IDE has looked at your efforts at writing code and found them to be acceptable. It tells you this by saying "Done Compiling" and reporting the size of the sketch to you: 450 bytes. The IDE is also telling you that the maximum size is 32,256 bytes, so you still have lots of room to make your sketch bigger.

Let's examine this boilerplate code that will form the starting point for every sketch that you ever write. There are some new things here. For example, there is the word **void** and some curly braces. Let's deal with **void** first.

Figure 3-4 *A sketch that will compile*

The line **void setup()** means that you are defining a function called **setup**. In Arduino, some functions are already defined for you, such as **digitalWrite** and **delay**, whereas you must or can define others for yourself. **setup** and **loop** are two functions that you must define for yourself in every sketch that you write.

The important thing to understand is that here you are not calling **setup** or **loop** like you would call **digitalWrite**, but you are actually creating these functions so that the Arduino system itself can call them. This is a difficult concept to grasp, but one way to think of it is as being similar to a definition in a legal document.

Most legal documents have a "definitions" section that might say, for example, something like the following:

```
Definitions.
The Author: The person or persons responsible for
creating the book
```

By defining a term in this way—for example, simply using the word "author" as shorthand for "The person or persons responsible for creating the book"—lawyers can make their documents shorter and more readable. Functions work much like such definitions. You define a function that you or the system itself can then use elsewhere in your sketches.

Going back to **void**, these two functions (**setup** and **loop**) do not return a value as some functions do, so you have to say that they are void, using the **void** keyword. If you imagine a function called **sin** that performed the trigonometric function of that name, then this function would return a value. The value returned to use from the call would be the sin of the angle passed as its argument.

Rather like a legal definition uses words to define a term, we write functions in C that can then be called from C.

After the special keyword **void** comes the name of the function and then parentheses to contain any arguments. In this case, there are no arguments, but we still have to include the parentheses there. There is no semicolon after the closing parenthesis because we are defining a function rather than calling it, so we need to say what will happen when something does call the function.

Those things that are to happen when the function is called must be placed between curly braces. Curly braces and the code in between them are known as a *block* of code, and this is a concept that you will meet again later.

Note that although you do have to define both the functions **setup** and **loop**, you do not actually have to put any lines of code in them. However, failing to add code will make your sketch a little dull.

Blink—Again!

The reason that Arduino has the two functions **setup** and **loop** is to separate the things that only need to be done once, when the Arduino starts running its sketch, from the things that have to keep happening continuously.

The function **setup** will just be run once when the sketch starts. Let's add some code to it that will blink the LED built onto the board. Add the lines to your sketch so that it appears as follows and then upload them to your board:

```
void setup()
{
  pinMode(13, OUTPUT);
  digitalWrite(13, HIGH);
}

void loop()
{
}
```

The **setup** function itself calls two built-in functions, **pinMode** and **digitalWrite**. You already know about **digitalWrite**, but **pinMode** is new. The function **pinMode** sets a particular pin to be either an input or an output. So, turning the LED on is actually a two-stage process. First, you have to set pin 13 to be an output, and second, you need to set that output to be high (5V).

When you run this sketch, on your board you will see that the LED comes on and stays on. This is not very exciting, so let's at least try to make it flash by turning it on and off in the **loop** function rather than in the **setup** function.

You can leave the **pinMode** call in the **setup** function because you only need to call it once. The project would still work if you moved it into the loop, but there is no need and it is a good programming habit to do things only once if you only need to do them once. So modify your sketch so that it looks like this:

```
void setup()
{
  pinMode(13, OUTPUT);
}

void loop()
{
 digitalWrite(13, HIGH);
 delay(500);
 digitalWrite(13, LOW);
}
```

Run this sketch and see what happens. It may not be quite what you were expecting. The LED is basically on all the time. Hmm, why should this be?

Try stepping through the sketch a line at a time in your head:

1. Run **setup** and set pin 13 to be an output.

2. Run **loop** and set pin 13 to high (LED on).

3. Delay for half a second.

4. Set pin 13 to low (LED off).

5. Run **loop** again, going back to step 2, and set pin 13 to high (LED on).

The problem lies between steps 4 and 5. What is happening is that the LED is being turned off, but the very next thing that happens is that it gets turned on again. This happens so quickly that it appears that the LED is on all the time.

The microcontroller chip on the Arduino can perform 16 million instructions per second. That's not 16 million C language commands, but it is still very fast. So, our LED will only be off for a few millionths of a second.

To fix the problem, you need to add another delay after you turn the LED off. Your code should now look like this:

```
// sketch 3-01
void setup()
{
  pinMode(13, OUTPUT);
}

void loop()
{
 digitalWrite(13, HIGH);
 delay(500);
 digitalWrite(13, LOW);
 delay(500);
}
```

Try again and your LED should blink away merrily once per second.

You may have noticed the comment at the top of the listing saying "sketch 3-01." To save you some typing, we have uploaded to the book's website all the sketches with such a comment at the top. You can download them from http://www.arduinobook.com.

Variables

In this Blink example, you use pin 13 and have to refer to it in three places. If you decided to use a different pin, then you would have to change the code in three places. Similarly, if you wanted to change the rate of blinking, controlled by the argument to delay, you would have to change 500 to some other number in more than one place.

Variables can be thought of as giving a name to a number. Actually, they can be a lot more powerful than this, but for now, you will use them for this purpose.

When defining a variable in C, you have to specify the type of the variable. We want our variables to be whole numbers, which in C are called **ints**. So to define a variable called **ledPin** with a value of 13, you need to write the following:

```
int ledPin 13;
```

Notice that because **ledPin** is a name, the same rules apply as those of function names. So, there cannot be any spaces. The convention is to start variables with a lowercase letter and begin each new word with an uppercase letter. Programmers will often call this "bumpy case" or "camel case."

Let's fit this into your Blink sketch as follows:

```
// sketch 3-02
int ledPin = 13;
int delayPeriod = 500;

void setup()
{
   pinMode(ledPin, OUTPUT);
}

void loop()
{
  digitalWrite(ledPin, HIGH);
  delay(delayPeriod);
  digitalWrite(ledPin, LOW);
  delay(delayPeriod);
}
```

We have also sneaked in another variable called **delayPeriod**.

Everywhere in the sketch where you used to refer to 13, you now refer to **ledPin**, and everywhere you used to refer to 500, you now refer to **delayPeriod**.

If you want to make the sketch blink faster, you can just change the value of **delayPeriod** in one place. Try changing it to 100 and running the sketch on your Arduino board.

There are other cunning things that you can do with variables. Let's modify your sketch so that the blinking starts really fast and gradually gets slower and slower, as if the Arduino is getting tired. To do this, all you need to do is to add something to the **delayPeriod** variable each time that you do a blink.

Modify the sketch by adding the single line at the end of the **loop** function so that it appears, as in the following listing, and then run the sketch on the Arduino board. Press the Reset button and see it start from a fast rate of flashing again.

```
sketch 3-03
int ledPin = 13;
int delayPeriod = 100;

void setup()
{
  pinMode(ledPin, OUTPUT);
}

void loop()
{
 digitalWrite(ledPin, HIGH);
 delay(delayPeriod);
 digitalWrite(ledPin, LOW);
 delay(delayPeriod);
 delayPeriod = delayPeriod + 100;
}
```

Your Arduino is doing arithmetic now. Every time that **loop** is called, it will do the normal flash of the LED, but then it will add 100 to the variable **delayPeriod**. We will come back to arithmetic shortly, but first you need a better way than a flashing LED to see what the Arduino is up to.

Experiments in C

You need a way to test your experiments in C. One way is to put the C that you want to test out into the **setup** function, evaluate them on the Arduino, and then have the Arduino display any output back to something called the Serial Monitor, as shown in Figures 3-5 and 3-6.

The Serial Monitor is part of the Arduino IDE. You access it by clicking on the rightmost icon in the toolbar. Its purpose is to act as a communication channel between your computer and the Arduino. You can type a message in the text entry area at the top of the Serial Monitor and when you press Return or click Send, it will send that message to the Arduino. Also if the Arduino has anything to say, this message will appear in the Serial Monitor. In both cases, the information is sent through the USB link.

Figure 3-5   *Writing C in* setup

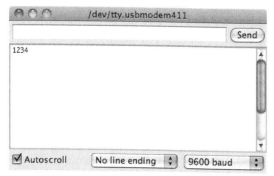

Figure 3-6   *The Serial Monitor*

As you would expect, there is a built-in function that you can use in your sketches to send a message back to the Serial Monitor. It is called **Serial.println** and it expects a single argument, which consists of the information that you want to send. This information is usually a variable.

You will use this mechanism to test out a few things that you can do with variables and arithmetic in C; frankly, it's the only way you can see the results of your experiments in C.

Numeric Variables and Arithmetic

The last thing you did was add the following line to your blinking sketch to increase the blinking period steadily:

```
delayPeriod = delayPeriod + 100;
```

Looking closely at this line, it consists of a variable name, then an equals sign, then what is called an expression (**delayPeriod + 100**). The equals sign does something called assignment. That is, it assigns a new value to a variable, and the value it is given is determined by what comes after the equals sign and before the semicolon. In this case, the new value to be given to the **delayPeriod** variable is the old value of **delayPeriod** plus 100.

Let's test out this new mechanism to see what the Arduino is up to by entering the following sketch, running it, and opening the Serial Monitor:

```
// sketch 3-04
void setup()
{
  Serial.begin(9600);
  int a = 2;
  int b = 2;
  int c = a + b;
  Serial.println(c);
}
void loop()
{}
```

Figure 3-7 shows what you should see in the Serial Monitor after this code runs.

To take a slightly more complex example, the formula for converting a temperature in degrees Centigrade into degrees Fahrenheit is to multiply

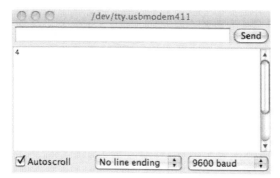

Figure 3-7　*Simple arithmetic*

it by 5, divide by 9, and then add 32. So you could write that in a sketch like this:

```
// sketch 3-05
void setup()
{
  Serial.begin(9600);
  int degC = 20;
  int degF;
  degF = degC * 9 / 5 + 32;
  Serial.println(degF);
}
void loop()
{}
```

There are a few things to notice here. First, note the following line:

```
int degC = 20;
```

When we write such a line, we are actually doing two things: We are declaring an **int** variable called **degC**, and we are saying that its initial value will be 20. Alternatively, you could separate these two things and write the following:

```
int degC;
degC = 20;
```

You must declare any variable just once, essentially telling the compiler what type of variable it is—in this case, **int**. However, you can assign the variable a value as many times as you want:

```
int degC;
degC = 20;
degC = 30;
```

So, in the Centigrade to Fahrenheit example, you are defining the variable **degC** and giving it an initial value of 20, but when you define **degF**, it does not get an initial value. Its value gets assigned on the next line, according to the conversion formula, before being sent to the Serial Monitor for you to see.

Looking at the expression, you can see that you use the asterisk (*) for multiplication and the slash (/) for division. The arithmetic operators +, −, *, and / have an order of precedence—that is, multiplications are done first, then divisions, then additions and subtractions. This is in accordance with the usual use of arithmetic. However, sometimes it makes it clearer to use parentheses in the expressions. So, for example, you could write the following:

```
degF = ((degC * 9) / 5) + 32;
```

The expressions that you write can be as long and complex as you need them to be, and in addition to the usual arithmetic operators, there are other less commonly used operators and a big collection of various mathematical functions that are available to you. You will learn about these later.

Commands

The C language has a number of built-in commands. In this section, we explore some of these and see how they can be of use in your sketches.

if

In our sketches so far, we have assumed that your lines of programming will be executed in order one after the other, with no exceptions. But what if you don't want to do that? What if you only want to execute part of a sketch if some condition is true?

Let's return to our gradually slowing-down Blinking LED example. At the moment, it will gradually get slower and slower until each blink is lasting hours. Let's look at how we can change it so that once it has slowed down to a certain point, it goes back to its fast starting speed.

To do this, you must use an **if** command; the modified sketch is as follows. Try it out.

```
// sketch 3-06
int ledPin = 13;
int delayPeriod = 100;

void setup()
{
  pinMode(ledPin, OUTPUT);
}

void loop()
{
  digitalWrite(ledPin, HIGH);
  delay(delayPeriod);
  digitalWrite(ledPin, LOW);
  delay(delayPeriod);
  delayPeriod = delayPeriod + 100;
  if (delayPeriod > 3000)
  {
    delayPeriod = 100;
  }
}
```

The **if** command looks a little like a function definition, but this resemblance is only superficial. The word in the parenthesis is not an argument; it is what is called *a condition*. So in this case, the condition is that the variable **delayPeriod** has a value that is greater than 3,000. If this is true, then the commands inside the curly braces will be executed. In this case, the code sets the value of **delayPeriod** back to 100.

If the condition is not true, then the Arduino will just continue on with the next thing. In this case, there is nothing after the "if", so the Arduino will run the **loop** function again.

Running through the sequence of events in your head will help you understand what is going on. So, here is what happens:

1. Arduino runs **setup** and initializes the LED pin to be an output.
2. Arduino starts running **loop**.
3. The LED turns on.
4. A delay occurs.
5. The LED turns off.
6. A delay occurs.
7. Add 100 to the **delayPeriod**.
8. If the delay period is greater than 3,000 set it back to 100.
9. Go back to step 2.

We used the symbol <, which means less than. It is one example of what are called comparison operators. These operators are summarized in the following table:

Operator	Meaning	Examples	Result
<	Less than	9 < 10 10 < 10	true false
>	Greater than	10 > 10 10 > 9	false true
<=	Less than or equal to	9 <= 10 10 <= 10	true true
>=	Greater than or equal to	10 >= 10 10 >= 9	true true
==	Equal to	9 == 9	true
!=	Not equal to	9 != 9	false

To compare two numbers, you use the == command. This double equals sign is easily confused with the character =, which is used to assign values to variables.

There is another form of **if** that allows you to do one thing if the condition is true and another if it is false. We will use this in some practical examples later in the book.

for

In addition to executing different commands under different circumstances, you also often will want to run a series of commands a number of times in a program. You already know one way of doing this, using the **loop** function. As soon as all the commands in the **loop** function have been run, it will start again automatically. However, sometimes you need more control than that.

So, for example, let's say that you want to write a sketch that blinks 20 times, then paused for 3 seconds, and then started again. You could do that by just repeating the same code over and over again in your **loop** function, like this:

```
// sketch 3-07
int ledPin = 13;
int delayPeriod = 100;

void setup()
{
  pinMode(ledPin, OUTPUT);
}

void loop()
{
 digitalWrite(ledPin, HIGH);
 delay(delayPeriod);
 digitalWrite(ledPin, LOW);
 delay(delayPeriod);

 digitalWrite(ledPin, HIGH);
 delay(delayPeriod);
 digitalWrite(ledPin, LOW);
 delay(delayPeriod);

 digitalWrite(ledPin, HIGH);
 delay(delayPeriod);
 digitalWrite(ledPin, LOW);
 delay(delayPeriod);
// repeat the above 4 lines another 17 times

 delay(3000);
}
```

But this requires a lot of typing and there are several much better ways to do this. Let's start by looking at how you can use a **for** loop and then look at another way of doing it using a counter and an **if** statement.

The sketch to accomplish this with a **for** loop is, as you can see, a lot shorter and easier to maintain than the previous example:

```
// sketch 3-08
int ledPin = 13;
int delayPeriod = 100;

void setup()
{
  pinMode(ledPin, OUTPUT);
}

void loop()
{
  for (int i = 0; i < 20; i ++)
  {
   digitalWrite(ledPin, HIGH);
   delay(delayPeriod);
   digitalWrite(ledPin, LOW);
   delay(delayPeriod);
  }
 delay(3000);
}
```

The **for** loop looks a bit like a function that takes three arguments, although here those arguments are separated by semicolons rather than the usual commas. This is just a quirk of the C language. The compiler will soon tell you when you get it wrong.

The first thing in the parentheses after **for** is a variable declaration. This specifies a variable to be used as a counter variable and gives it an initial value—in this case, 0.

The second part is a condition that must be true for you to stay in the **loop**. In this case, you will stay in the **loop** as long as **i** is less than 20, but as soon as **i** is 20 or more, the program will stop doing the things inside the **loop**.

The final part is what to do every time you have done all the things in the **loop**. In this case, that is to increment **i** by 1 so that it can, after 20 trips around the **loop**, cease to be less than 100 and cause the program to exit the **loop**.

Try entering this code and running it. The only way to get familiar with the syntax and all that pesky punctuation is to type it in and have the compiler tell you when you have done something wrong. Eventually it will all start to make sense.

One potential downside of this approach is that the **loop** function is going to take a long time. This is not a problem for this sketch, because all it is doing is flashing an LED. But often, the **loop** function in a sketch will also be checking that keys have been pressed or that serial communications have been received. If the processor is busy inside a **for** loop, it will not be able to do this. Generally, it is a good idea to make the **loop** function run as fast as possible so that it can be run as frequently as possible.

The following sketch shows how to achieve this:

```
// sketch 3-09
int ledPin = 13;
int delayPeriod = 100;
int count = 0;
void setup()
{
  pinMode(ledPin, OUTPUT);
}

void loop()
{
 digitalWrite(ledPin, HIGH);
 delay(delayPeriod);
 digitalWrite(ledPin, LOW);
 delay(delayPeriod);
 count ++;
 if (count == 20)
 {
   count = 0;
   delay(3000);
 }
}
```

You may have noticed the following line:

```
count ++;
```

This is just C shorthand for the following:

```
count = count + 1;
```

So now each time that **loop** is run, it will take just a bit more than 200 milliseconds, unless it's the 20th time round the loop, in which case it will take the same plus the three seconds delay between each batch of 20 flashes. In fact, for some applications, even this is too slow, and purists would say that you should not use **delay** at all. The best solution depends on the application.

while

Another way of looping in C is to use the **while** command in place of the **for** command. You can accomplish the same thing as the preceding **for** example using a **while** command as follows:

```
int i = 0;
while (i < 20)
{
  digitalWrite(ledPin, HIGH);
  delay(delayPeriod);
  digitalWrite(ledPin, LOW);
  delay(delayPeriod);
  i ++;
}
```

The expression in parentheses after **while** must be true to stay in the **loop**. When it is no longer true, then the sketch continues running the commands after the final curly brace.

The #define Directive

For constant values like pin assignments that do not change during the running of a sketch, there is an alternative to using a variable. You can use a command called **#define** that allows you to associate a value with a name. Everywhere that this name appears in your sketch, the value will be substituted before the sketch is compiled.

As an example, you could define a pin assignment for a LED like this:

```
#define ledPin 13
```

Note that the **#define** directive does not use an "=" between the name and the value. It does not even need a ";" on the end. This is because it is not actually part of the C language itself; but is called a pre-compiler directive that is run before compilation.

This approach, in the author's opinion, is less easy to read than using a variable, but it does have the advantage that you do not use any memory to store it. It is something to consider if memory is at a premium.

Conclusion

This chapter has got you started with C. You can make LEDs blink in various exciting ways and get the Arduino to send results back to you over the USB by using the **Serial.println** function. You also worked out how to use **if** and **for** commands to control the order in which your commands are executed, and learned a little about making an Arduino do some arithmetic.

In the next chapter, you will look more closely at functions. The chapter will also introduce the variable types other than the **int** type that you used in this chapter.

4

Functions

This chapter focuses mostly on the type of functions that you can write yourself rather than the built-in functions such as **digitalWrite** and **delay** that are already defined for you.

The reason that you need to be able to write your own functions is that as sketches start to get a little complicated, then your **setup** and **loop** functions will grow and grow until they are long and complicated and it becomes difficult to see how they work.

The biggest problem in software development of any sort is managing complexity. The best programmers write software that is easy to look at and understand and requires very little in the way of explanation.

Functions are a key tool in creating easy-to-understand sketches that can be changed without difficulty or risk of the whole thing falling into a crumpled mess.

What Is a Function?

A function is a little like a program within a program. You can use it to wrap up some little thing that you want to do. A function that you define can be called from anywhere in your sketch and contains its own variables and its own list of commands. When the commands have been run, execution returns to the point just after wherever it was in the code that called the function.

By way of an example, code that flashes a light-emitting diode (LED) is a prime example of some code that should be put in a function. So let's modify our basic "blink 20 times" sketch to use a function that we will create called **flash**:

```
// sketch 4-01
int ledPin = 13;
int delayPeriod = 250;

void setup()
{
  pinMode(ledPin, OUTPUT);
}

void loop()
{
  for (int i = 0; i < 20; i ++)
  {
    flash();
  }
  delay(3000);
}

void flash()
{
   digitalWrite(ledPin, HIGH);
   delay(delayPeriod);
   digitalWrite(ledPin, LOW);
   delay(delayPeriod);
}
```

So, all we have really done here is to move the four lines of code that flash the LED from the middle of the **for** loop to be in a function of their own called **flash**. Now you can make the LED flash any time you like by just calling the new function by writing **flash()**. Note the empty parentheses after the function name. This indicates that the function does not take any parameters. The delay value that it uses is set by the same **delayPeriod** function that you used before.

Parameters

When dividing your sketch up into functions, it is often worth thinking about what service a function could provide. In the case of **flash**, this is fairly obvious. But this time, let's give this function parameters that tell it both, how many times to flash and how short or long the flashes should be. Read through the following code and then I will explain just how parameters work in a little more detail.

```
// sketch 4-02
int ledPin = 13;
int delayPeriod = 250;

void setup()
{
  pinMode(ledPin, OUTPUT);
}

void loop()
{
  flash(20, delayPeriod);
  delay(3000);
}

void flash(int numFlashes, int d)
{
  for (int i = 0; i < numFlashes; i ++)
  {
    digitalWrite(ledPin, HIGH);
    delay(d);
    digitalWrite(ledPin, LOW);
    delay(d);
  }
}
```

Now, if we look at our **loop** function, it has only two lines in it. We have moved the bulk of the work off to the **flash** function. Notice how when we call **flash** we now supply it with two arguments in parentheses.

Where we define the function at the bottom of the sketch, we have to declare the type of variable in the parameters. In this case, they are both **int**s. We are in fact defining new variables. However, these variables (**numFlashes** and **d**) can only be used within the **flash** function.

This is a good function because it wraps up everything you need in order to flash an LED. The only information that it needs from outside of the function is to which pin the LED is attached. If you wanted, you could make this a parameter too—something that would be well worth doing if you had more than one LED attached to your Arduino.

Global, Local, and Static Variables

As was mentioned before, parameters to a function can be used only inside that function. So, if you wrote the following code, you would get an error:

```
void indicate(int x)
{
   flash(x, 10);
}
x = 15;
```

On the other hand, suppose you wrote this:

```
int x;
void indicate(int x)
{
   flash(x, 10);
}
x = 15;
```

This code would not result in a compilation error. However, you need to be careful, because you now actually have two variables called **x** and they can each have different values. The one that you declared on the first line is called a *global variable*. It is called *global* because it can be used anywhere you like in the program, including inside any functions.

However, because you use the same variable name **x** inside the function, as a parameter, you cannot use the global variable **x** simply because whenever you refer to **x** inside the function, the "local" version of **x** has priority. The parameter **x** is said to shadow the global variable of the same name. This can lead to some confusion when trying to debug a project.

In addition to defining parameters, you can also define variables that are not parameters but are just for use within a function. These are called *local variables*. For example:

```
void indicate(int x)
{
  int timesToFlash = x * 2;
  flash(timesToFlash, 10);
}
```

The local variable **timesToFlash** will only exist while the function is running. As soon as the function has finished its last command, it will disappear. This means that local variables are not accessible from anywhere in your program other than in the function in which they are defined.

So, for instance, the following example will cause an error:

```
void indicate(int x)
{
  int timesToFlash = x * 2;
  flash(timesToFlash, 10);
}
timesToFlash = 15;
```

Seasoned programmers generally treat global variables with suspicion. The reason is that they go against the principal of encapsulation. The idea of *encapsulation* is that you should wrap up in a package everything that has to do with a particular feature. Hence functions are great for encapsulation. The problem with "globals" (as global variables are often called) is that they generally get defined at the beginning of a sketch and may then be used all over the sketch. Sometimes there is a perfectly legitimate reason for this. Other times, people use them in a lazy way when it would be far more appropriate to pass parameters. In our examples so far, **ledPin** is a good use of a global variable. It's also very convenient and easy to find up at the top of the sketch, making it easy to change. Actually, **ledPin** is really a constant, because although you may change it and then recompile your sketch, you are unlikely to allow the variable to change while the sketch is actually running. For this reason, you may prefer to use the **#define** command we described in Chapter 3.

Another feature of local variables is that their value is initialized every time the function is run. This is nowhere more true (and often inconvenient) than in the **loop** function of an Arduino sketch. Let's try and use a local variable in place of global variable in one of the examples from the previous chapter:

```
// sketch 4-03
int ledPin = 13;
int delayPeriod = 250;
void setup()
{
  pinMode(ledPin, OUTPUT);
}

void loop()
{
 int count = 0;
 digitalWrite(ledPin, HIGH);
 delay(delayPeriod);
 digitalWrite(ledPin, LOW);
 delay(delayPeriod);
 count ++;
 if (count == 20)
 {
   count = 0;
   delay(3000);
 }
}
```

Sketch 4-03 is based on the sketch 3-09, but attempts to use a local variable instead of the global variable to count the number of flashes.

This sketch is broken. It will not work, because every time **loop** is run, the variable **count** will be given the value 0 again, so **count** will never reach 20 and the LED will just keep flashing forever. The whole reason that we made **count** a global in the first place was so that its value would not be reset. The only place that we use **count** is in the **loop** function, so this is where it should be placed.

Fortunately, there is a mechanism in C that gets around this conundrum. It is the keyword **static**. When you use the keyword **static** in front of a variable declaration in a function, it has the effect of initializing the variable only the first time that the function is run. Perfect! That's just what is required in this situation. We can keep our variable in the function where it's used without it getting set back to 0 every time the function runs. Sketch 4-04 shows this in operation:

```
// sketch 4-04
int ledPin = 13;
int delayPeriod = 250;
void setup()
{
  pinMode(ledPin, OUTPUT);
}

void loop()
{
  static int count = 0;
  digitalWrite(ledPin, HIGH);
  delay(delayPeriod);
  digitalWrite(ledPin, LOW);
  delay(delayPeriod);
  count ++;
  if (count == 20)
  {
    count = 0;
    delay(3000);
  }
}
```

Return Values

Computer science, as an academic discipline, has as its parents mathematics and engineering. This heritage lingers on in many of the names associated with programming. The word *function* is itself a mathematical term. In mathematics, the input to the function (the argument) completely determines the output. We have written functions that take an input, but none that give us back a value. All our functions have been "void" functions. If a function returns a value, then you specify a return type.

Let's look at writing a function that takes a temperature in degrees Centigrade and returns the equivalent in degrees Fahrenheit:

```
int centToFaren(int c)
{
  int f = c * 9 / 5 + 32;
  return f;
}
```

The function definition now starts with **int** rather than **void** to indicate that the function will return an **int** to whatever calls it. This might be a bit of code that looks like this:

```
int pleasantTemp = centToFaren(20);
```

Any non-void function has to have a **return** statement in it. If you do not put one in, the compiler will tell you that it is missing. You can have more than one **return** in the same function. This might arise if you have an **if** statement with alternative actions based on some condition. Some programmers frown on this, but if your functions are small (as all functions should be), then this practice will not be a problem.

The value after **return** can be an expression; it does not have to just be the name of a variable. So you could compress the preceding example into the following:

```
int centToFaren(int c)
{
  return (f = c * 9 / 5 + 32);
}
```

If the expression being returned is more than just a variable name, then it should be enclosed in parentheses as in the preceding example.

Other Variable Types

All our examples of variables so far have been **int** variables. This is by far the most commonly used variable type, but there are some others that you should be aware of.

floats

One such type, which is relevant to the previous temperature conversion example, is **float**. This variable type represents floating point numbers— that is, numbers that may have a decimal point in them, such as 1.23. You need this variable type when whole numbers are just not precise enough. Note the following formula:

```
f = c * 9 / 5 + 32
```

If you give **c** the value 17, then **f** will be 17 * 9 / 5 + 32 or 62.6. But if **f** is an **int**, then the value will be truncated to 62.

The problem becomes even worse if we are not careful of the order in which we evaluate things. For instance, suppose that we did the division first, as follows:

```
f = (c / 5) * 9 + 32
```

Then in normal math terms, the result would still be 62.6, but if all the numbers are **int**s, then the calculation would proceed as follows:

1. 17 is divided by 5, which gives 3.4, which is then truncated to 3.

2. 3 is then multiplied by 9 and 32 is added to give a result of 59—which is quite a long way from 62.6.

For circumstances like this, we can use **floats**. In the following example, our temperature conversion function is rewritten to use **floats**:

```
float centToFaren(float c)
{
   float f = c * 9.0 / 5.0 + 32.0;
   return f;
}
```

Notice how we have added .0 to the end of our constants. This ensures that the compiler knows to treat them as **floats** rather than **ints**.

boolean

Boolean values are logical. They have a value that is either true or false.

In the C language, *Boolean* is spelled with a lowercase *b*, but in general use, *Boolean* has an uppercase initial letter, as it is named after the mathematician George Boole, who invented the Boolean logic that is crucial to computer science.

You may not realize it, but you have already met Boolean values when we were looking at the **if** command. The condition in an **if** statement, such as **(count == 20)**, is actually an expression that yields a **boolean** result. The operator **==** is called a comparison operator. Whereas **+** is an arithmetic operator that adds two numbers together, **==** is a comparison operator that compares two numbers and returns a value of either true or false.

You can define Boolean variables and use them as follows:

```
boolean tooBig = (x > 10);
if (tooBig)
{
  x = 5;
}
```

Boolean values can be manipulated using Boolean operators. So, similar to how you can perform arithmetic on numbers, you can also perform operations on Boolean values. The most commonly used Boolean operators are **and**, which is written as **&&**, and **or**, which is written as **||**.

Figure 4-1 shows truth tables for the **and** and **or** operators.

From the truth tables in Figure 4-1, you can see that for **and**, if both A and B are true, then the result will be true; otherwise, the result will be false.

On the other hand, with the **or** operator, if either A or B or both A and B are true, then the result will be true. The result will be false only if neither A nor B is true.

In addition to **and** and **or**, there is the **not** operator, written as **!**. You will not be surprised to learn that "not true" is false and "not false" is true.

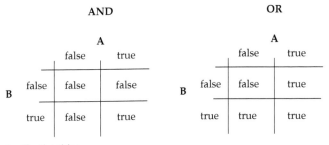

Figure 4-1 *Truth tables*

You can combine these operators into Boolean expressions in your **if** statements, as the following example illustrates:

```
if ((x > 10) && (x < 50))
```

Other Data Types

As you have seen, the **int** and occasionally the **float** data types are fine for most situations; however, some other numeric types can be useful under some circumstances. In an Arduino sketch, the **int** type uses 16 bits (binary digits). This allows it to represent numbers between –32767 and 32768.

Other data types available to you are summarized in Table 4-1. This table is provided mainly for reference. You will use some of these other types as you progress through the book.

Type	Memory (bytes)	Range	Notes
boolean	1	true or false (0 or 1)	
char	1	–128 to +128	Used to represent an American Standard Code for Information Interchange (ASCII) character code; e.g., *A* is represented as 65. Its negative numbers are not normally used.
byte	1	0 to 255	Often used for communicating serial data, as a single unit of data. See Chapter 9.
int	2	–32768 to +32767	
unsigned int	2	0 to 65536	Can be used for extra precision where negative numbers are not needed. Use with caution, as arithmetic with **int**s may cause unexpected results.
long	4	2,147,483,648 to 2,147,483,647	Needed only for representing very big numbers.
unsigned long	4	0 to 4,294,967,295	See **unsigned int**.
float	4	–3.4028235E+38 to + 3.4028235E+38	
double	4	Same as **float**	Normally, this would be 8 bytes and higher precision than **float** with a greater range. However, on Arduino, it is the same as **float**.

Table 4-1 *Data Types in C*

One thing to consider is that if data types exceed their range, then strange things happen. So, if you have a byte variable with 255 in it and you add 1 to it, you get 0. More alarmingly, if you have an **int** variable with 32767 and you add 1 to it, you will end up with –32768.

Until you are completely comfortable with these different data types, I would recommend sticking to **int**, as it works for pretty much everything.

Coding Style

The C compiler does not really care about how you lay out your code. For all it cares, you can write everything on a single line with semicolons between each statement. However, well-laid-out, neat code is much easier to read and maintain than poorly laid-out code. In this sense, reading code is just like reading a book: Formatting is important.

To some extent, formatting is a matter of personal taste. No one likes to think that he has bad taste, so arguments about how code should look can become personal. It is not unknown for programmers, on being required to do something with someone else's code, to start by reformatting all the code into their preferred style of presentation.

As an answer to this problem, coding standards are often laid down to encourage everyone to lay out his or her code in the same way and adopt "good practice" when writing programs.

The C language has a de facto standard that has evolved over the years, and this book is generally faithful to that standard.

Indentation

In the example sketches that you have seen, you can see that we often indent the program code from the left margin. So, for example when defining a **void** function, the **void** keyword is at the left margin, as is the opening curly brace on the next line, but then all the text within the curly braces is indented. The amount of indentation does not really matter. Some people use two spaces, some four. You can also press TAB to indent. In this book, we use two spaces for indentation.

If you were to have an `if` statement inside a function definition, then once again you would add two more spaces for the lines within the curly braces of the **if** command, as in the following example:

```
void loop()
{
  static int count = 0;
  count ++;
  if (count == 20)
  {
    count = 0;
    delay(3000);
  }
}
```

You might include another **if** inside the first **if**, which would add yet another level of indentation, making six spaces from the left margin.

All of this might sound a bit trivial, but if you ever sort through someone else's badly formatted sketches, you will find it very difficult.

Opening Braces

There are two schools of thought as to where to put the first curly brace in a function definition, **if** statement, or **for** loop. One way is to place the curly brace on the line after the rest of the command, as we have in all the examples so far, or put it on the same line, like this:

```
void loop() {
  static int count = 0;
  count ++;
  if (count == 20) {
    count = 0;
    delay(3000);
  }
}
```

This style is most commonly used in the Java programming language, which shares much of the same syntax as C. I prefer the first form, which seems to be the form most commonly used in the Arduino world.

Whitespace

The compiler ignores spaces tabs and new lines, apart from using them as a way of separating the "tokens" or words in your sketch. Thus the following example will compile without a problem:

```
void loop() {static int
count=0;count++;if(
count==20){count=0;
delay(3000);}}
```

This will work, but good luck trying to read it.

Where assignments are made, some people will write the following:

```
int a = 10;
```

But others will write the following:

```
int a=10;
```

Which of these two styles you use really does not matter, but it is a good idea to be consistent. I use the first form.

Comments

Comments are text that is kept in your sketch along with all the real program code, but which actually performs no programming function whatsoever. The sole purpose of comments is to be a reminder to you or others as to why the code is written as it is. A comment line may also be used to present a title.

The compiler will completely ignore any text that is marked as being a comment. We have included comments as titles at the top of many of the sketches in the book so far.

There are two forms of syntax for comments:

- The single line comment that starts with // and finishes at the end of the line
- The multiline comment that starts with a /* and ends with a */

The following is an example using both forms of comments:

```
/* A not very useful loop function.
Written by: Simon Monk
To illustrate the concept of comments
*/
void loop() {
  static int count = 0;
  count ++; // a single line comment
  if (count == 20) {
    count = 0;
    delay(3000);
  }
}
```

In this book, I mostly stick to the single-line comment format.

Good comments help explain what is happening in a sketch or how to use the sketch. They are useful if others are going to use your sketch, but equally useful to yourself when you are looking at a sketch that you have not worked on for a few weeks.

Some people are told in programming courses that the more comments, the better. Most seasoned programmers will tell you that well-written code requires very little in the way of comments because it is self-explanatory. You should use comments for the following reasons:

- To explain anything you have done that is a little tricky or not immediately obvious

- To describe anything that the user needs to do that is not part of the program; for example, **// this pin should be connected to the transistor controlling the relay**

- To leave yourself notes; for example, **// todo: tidy this - it's a mess**

This last point illustrates a useful technique of **todo**s in comments. Programmers often put **todo**s in their code to remind themselves of something they need to do later. They can always use the search facility in their integrated development environment (IDE) to find all occurrences of **// todo** in their program.

The following are *not* good examples of reasons you should use comments:

- To state the blatantly obvious; for example, **a = a + 1; // add 1 to a**.

- To explain badly written code. Don't comment on it; just write it clearly in the first place.

Conclusion

This has been a bit of a theoretical chapter. You have had to absorb some new abstract concepts concerned with organizing our sketches into functions and adopting a style of programming that will save you time in the long run.

In the next chapter, you can start to apply some of what you have learned and look at better ways of structuring your data and using text strings.

5

Arrays and Strings

After reading Chapter 4, you have a reasonable appreciation as to how to structure your sketches to make your life easier. If there is one thing that a good programmer likes, it's an easy life. Now our attention is going to turn to the data that you use in your sketches.

The book *Algorithms + Data Structures = Programs* by Niklaus Wirth has been around for a good while now, but still manages to capture the essences of computer science and programming in particular. I can strongly recommend it to anyone who finds themselves bitten by the programming bug. It also captures the idea that to write a good program, you need to think about both the algorithm (what you do) and the structure of the data you use.

You have looked at **loops**, **if** statements, and what is called the "algorithmic" side of programming an Arduino; you are now going to turn to how you structure your data.

Arrays

Arrays are a way of containing a list of values. The variables that you have met so far have contained only a single value, usually an **int**. By contrast, an array contains a list of values, and you can access any one of those values by its position in the list.

C, in common with the majority of programming languages, begins its index positions at 0 rather than 1. This means that the first element is actually element zero.

To illustrate the use of arrays, we could create an example application that repeatedly flashes "SOS" in Morse code using the Arduino board's built-in LED.

Morse code used to be a vital method of communication in the 19th and 20th centuries. Because of its coding of letters as a series of long and short dots, Morse code can be sent over telegraph wires, over a radio link, and using signaling lights. The letters "SOS" (an acronym for "save our souls") is still recognized as an international signal of distress.

The letter "S" is represented as three short flashes (dots) and the letter "O" by three long flashes (dashes). You are going to use an array of **int**s to hold the duration of each flash that you are going to make. You can then use a **for** loop to step through each of the items in the array, making a flash of the appropriate duration.

First let's have a look at how you are going to create an array of **int**s containing the durations.

```
int durations[] = {200, 200, 200, 500, 500, 500, 200, 200, 200};
```

You indicate that a variable contains an array by placing [] after the variable name.

In this case, you are going to set the values for the durations at the time that you create the array. The syntax for doing this is to use curly braces and then values each separated by commas. Don't forget the semicolon on the end of the line.

You can access any given element of the array using the square bracket notation. So, if you want to get the first element of the array, you can write the following:

```
durations[0]
```

To illustrate this, let's create an array and then print out all its values to the Serial Monitor:

```
// sketch 5-01
int ledPin = 13;

int durations[] = {200, 200, 200, 500, 500, 500, 200, 200, 200};
```

```
void setup()
{
  Serial.begin(9600);
  for (int i = 0; i < 9; i++)
  {
    Serial.println(durations[i]);
  }
}

void loop() {}
```

Upload the sketch to your board and then open the Serial Monitor. If all is well, you will see something like Figure 5-1.

This is quite neat, because if you wanted to add more durations to the array, all you would need to do is add them to the list inside the curly braces and change "9" in the **for** loop to the new size of the array.

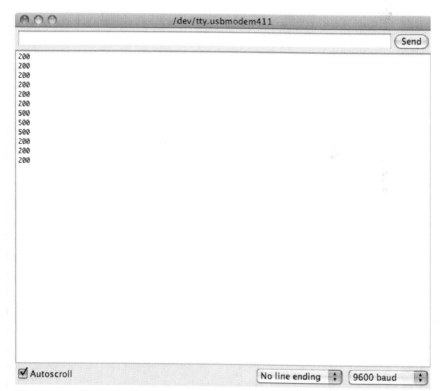

Figure 5-1 *The Serial Monitor Showing the Output of Sketch 5-01*

You have to be a little careful with arrays, because the compiler will not try and stop you from accessing elements of data that are beyond the end of the array. This is because the array is really a pointer to an address in memory, as shown in Figure 5-2.

Programs keep their data, both ordinary variables and arrays, in *memory*. Computer memory is arranged much more rigidly than the human kind of memory. It is easiest to think of the memory in an Arduino as a collection of pigeonholes. When you define an array of nine elements, for example, the next available nine pigeonholes are reserved for its use and the variable is said to point at the first pigeonhole or *element* of the array.

Going back to our point about access being allowed beyond the bounds of your array, if you decided to access **durations[10]**, then you would still

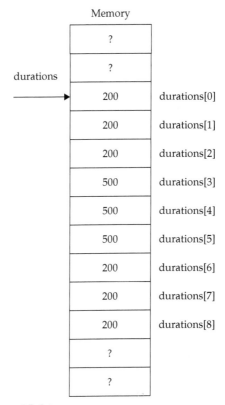

Figure 5-2   *Arrays and Pointers*

get back an **int**, but the value of this **int** could be anything. This is in itself fairly harmless, except that if you accidentally get a value outside of the array, you are likely to get confusing results in your sketch.

However, what is far worse is if you try changing a value outside of the size of the array. For instance, if you were to include something like the following in your program, the results could simply make your sketch break:

```
durations[10] = 0;
```

The pigeonhole **durations[10]** may be in use as some completely different variable. So always make sure that you do not go outside of the size of the array. If your sketch starts behaving strangely, then check for this kind of problem.

Morse Code SOS Using Arrays

Sketch 5-02 shows how you can use an array to make your emergency signal of SOS:

```
// sketch 5-02
int ledPin = 13;

int durations[] = {200, 200, 200, 500, 500, 500, 200, 200, 200};

void setup()
{
  pinMode(ledPin, OUTPUT);
}

void loop()
{
  for (int i = 0; i < 9; i++)
  {
    flash(durations[i]);
  }
  delay(1000);
}

void flash(int delayPeriod)
{
  digitalWrite(ledPin, HIGH);
  delay(delayPeriod);
  digitalWrite(ledPin, LOW);
  delay(delayPeriod);
}
```

An obvious advantage of this approach is that it is very easy to change the message by simply altering the **durations** array. In sketch 5-05, you will take the use of arrays a stage further to make a more general-purpose Morse code flasher.

String Arrays

In the programming world, the word *string* has nothing to do with long thin stuff that you tie knots in. A string is a sequence of characters. It's the way you can get your Arduino to deal with text. For example, the sketch 5-03 will repeatedly send the text "Hello" to the Serial Monitor one time per second:

```
// sketch 5-03
void setup()
{
  Serial.begin(9600);
}

void loop()
{
  Serial.println("Hello");
  delay(1000);
}
```

String Literals

String literals are enclosed in double quotation marks. They are literal in the sense that the string is a constant, rather like the **int** 123.

As you would expect, you can put strings in a variable. There is also an advanced string library, but for now you will use standard C strings, such as the one in sketch 5-03.

In C, a string literal is actually an array of the type **char**. The type **char** is a bit like **int** in that it is a number, but that number is between 0 and 127 and represents one character. The character may be a letter of the alphabet, a number, a punctuation mark, or a special character such as a tab or

Character	ASCII code (decimal)
a–z	97–122
A–Z	65–90
0–9	48–57
space	32

Table 5-1   *Common ASCII Codes*

a line feed. These number codes for letters use a standard called ASCII; Some of the most commonly used ASCII codes are shown in Table 5-1.

The string literal "Hello" is actually an array of characters, as shown in Figure 5-3.

Note that the string literal has a special null character at the end. This character is used to indicate the end of the string.

String Variables

As you would expect, string variables are very similar to array variables, except that there is a useful shorthand method for defining their initial value.

```
char name[] = "Hello";
```

This defines an array of characters and initializes it to the word "Hello." It will also add a final null value (ASCII 0) to mark the end of the string.

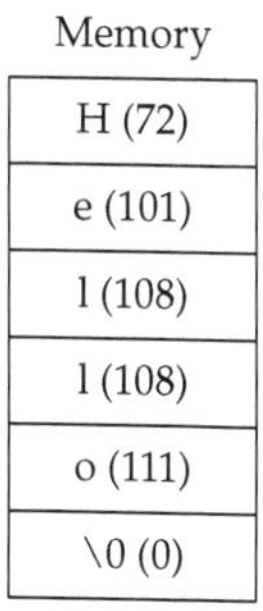

Memory

H (72)
e (101)
l (108)
l (108)
o (111)
\0 (0)

Figure 5-3   *The String Literal "Hello"*

Although the preceding example is most consistent with what you know about writing arrays, it would be more common to write the following:

```
char *name = "Hello";
```

This is equivalent, and the * indicates a pointer. The idea is that **name** points to the first **char** element of the **char** array. That is the memory location that contains the letter *H*.

You can rewrite sketch 5-03 to use a variable as well as a string constant, as follows:

```
// sketch 5-04
char message[] = "Hello";

void setup()
{
  Serial.begin(9600);
}

void loop()
{
  Serial.println(message);
  delay(1000);
}
```

A Morse Code Translator

Let's put together what you have learned about arrays and strings to build a more complex sketch that will accept any message from the Serial Monitor and flash it out on the built-in LED.

The letters in Morse code are shown in Table 5-2.

Some of the rules of Morse code are that a dash is three times as long as a dot, the time between each dash or dot is equal to the duration of a dot, the space between two letters is the same length as a dash, and the space between two words is the same duration as seven dots.

For this project, we will not worry about punctuation, although it would be an interesting exercise for you to try adding this to the sketch. For a full list of all the Morse characters, see en.wikipedia.org/wiki/Morse_code.

A	.-	N	-.	0	-----
B	-...	O	---	1	.----
C	-.-.	P	.--.	2	..---
D	-..	Q	--.-	3	...--
E	.	R	.-.	4	....-
F	..-.	S	...	5	.....
G	--.	T	-	6	-....
H	....	U	..-	7	--...
I	..	V	...-	8	---..
J	.---	W	.--	9	----.
K	-.-	X	-..-		
L	.-..	Y	-.--		
M	--	Z	--..		

Table 5-2 *Morse Code Letters*

Data

You are going to build this example a step at a time, starting with the data structure that you are going to use to represent the codes.

It is important to understand that there is no one solution to this problem. Different programmers will come up with different ways to solve it. So, it is a mistake to think to yourself, "I would never have come up with that." Well, no, quite possibly you would come up with something different and better. Everyone thinks in different ways, and this solution happens to be the one that first popped into the author's head.

Representing the data is all about finding a way of expressing Table 5-2 in C. In fact, you are going to split the data into two tables: one for the letters, and one for the numbers. The data structure for the letters is as follows:

```
char* letters[] = {
    ".-", "-...", "-.-.", "-..", ".", // A-I
    "..-.", "--.", "....", "..",
    ".---", "-.-", ".-..", "--", "-.", // J-R
    "---", ".--", "--.-", ".-.",
    "...", "-", "..-", "...-", ".--", // S-Z
    "-..-", "-.--", "--.."
};
```

What you have here is an array of string literals. So, because a string literal is actually an array of **char**, what you actually have here is an array of arrays—something that is perfectly legal and really quite useful.

This means that to find Morse for *A*, you would access **letters[0]**, which would give you the string .-. This approach is not terribly efficient, because you are using a whole byte (eight bits) of memory to represent a dash or a dot, which could be represented in a bit. However, you can easily justify this approach by saying that the total number of bytes is still only about 90 and we do have 2K to play with. Equally importantly, it makes the code easy to understand.

Numbers use the same approach:

```
char* numbers[] = {
    "-----", ".----", "..---", "...--", "....-",
    ".....", "-....", "--...", "---..", "----."};
```

Globals and Setup

You need to define a couple of global variables: one for the delay period for a dot, and one to define which pin the LED is attached to:

```
int dotDelay = 200;
int ledPin = 13;
```

The **setup** function is pretty simple; you just need to set the **ledPin** as an output and set up the serial port:

```
void setup()
{
  pinMode(ledPin, OUTPUT);
  Serial.begin(9600);
}
```

The *loop* function

You are now going to start on the real processing work in the **loop** function. The algorithm for this function is as follows:

- If there is a character to read from USB:
  - If it's a letter, flash it using the letters array

- If it's a number, flash it using the numbers array

- If it's a space, flash four times the dot delay

That's all. You should not think too far ahead. This algorithm represents what you want to do, or what your *intention* is, and this style of programming is called *programming by intention*.

If you write this algorithm in C, it will look like this:

```
void loop()
{
  char ch;
  if (Serial.available() > 0)
  {
    ch = Serial.read();
    if (ch >= 'a' && ch <= 'z')
    {
      flashSequence(letters[ch - 'a']);
    }
    else if (ch >= 'A' && ch <= 'Z')
    {
      flashSequence(letters[ch - 'A']);
    }
    else if (ch >= '0' && ch <= '9')
    {
      flashSequence(numbers[ch - '0']);
    }
    else if (ch == ' ')
    {
     delay(dotDelay * 4);       // gap between words
    }
  }
}
```

There are a few things here that need explaining. First, there is **Serial**
.available(). To understand this, you first need to know a little about how an Arduino communicates with your computer over USB. Figure 5-4 summarizes this process.

In the situation where the computer is sending data from the Serial Monitor to the Arduino board, then the USB is converted from the USB signal levels and protocol to something that the microcontroller on the Arduino board can use. This conversion happens in a special-purpose

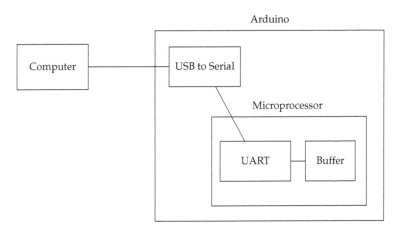

Figure 5-4 *Serial communication with an Arduino*

chip on the Arduino board. The data is then received by a part of the microcontroller called the Universal Asynchronous Receiver/Transmitter (UART). The UART places the data it receives into a buffer. The buffer is a special area of memory (128 bytes) that can hold data that is removed as soon as it is read.

This communication happens regardless of what your sketch is doing. So, even though you may be merrily flashing LEDs, data will still arrive in the buffer and sit there until you are ready to read it. You can think of the buffer as being a bit like an e-mail inbox.

The way that you check to see whether you "have mail" is to use the function **Serial.available()**. This function returns the number of bytes of data in the buffer that are waiting for you to read. If there are no messages waiting to be read, then the function returns 0. This is why the **if** statement checks to see that there are more than zero bytes available to read, and if they are, then the first thing that the statement does is read the next available **char**, using the function **Serial.read()**. This function gets assigned to the local variable **ch**.

Next is another **if** to decide what kind of thing it is that you want to flash:

```
if (ch >= 'a' && ch <= 'z')
{
   flashSequence(letters[ch - 'a']);
}
```

At first, this might seem a bit strange. You are using **<=** and **>=** to compare characters. You can do that because each character is actually represented by a number (its ASCII code). So, if the code for the character is somewhere between *a* and *z* (97 and 122), then you know that the character that has come from the computer is a lowercase letter. You then call a function that you have not written yet called **flashSequence**, to which you will pass a string of dots and dashes; for example, to flash *a*, you would pass it .- as its argument.

You are devolving responsibility to this function for actually doing the flashing. You are not trying to do it inside the **loop**. This lets us keep the code easy to read.

Here is the C that determines the string of dashes and dots that you need to send to the **flashSequence** function:

```
letters[ch - 'a']
```

Once again, this looks a little strange. The function appears to be subtracting one character from another. This is actually a perfectly reasonable thing to do, because the function is actually subtracting the ASCII values.

Remember that you are storing the codes for the letters in an array. So the first element of the array contains a string of dashes and dots for the letter *A*, the second element includes the dots and dashes for *B*, and so on. So you need to find the right position in the array for the letter that you have just fetched from the buffer. The position for any lowercase letter will be the character code for the letter minus the character code for *a*. So, for example, $a - a$ is actually $97 - 97 = 0$. Similarly, $c - a$ is actually $99 - 97 = 2$. So, in the following statement, if ch **is** the letter *c*, then the bit inside the square brackets would evaluate to 2, and you would get element 2 from the array, which is -.-:

What this section has just described is concerned with lowercase letters. You also have to deal with uppercase letters and numbers. These are both handled in a similar manner.

The *flashSequence* Function

We have assumed a function called **flashSequence** and made use of it, but now you need to write it. We have planned for it to take a string containing a series of dashes and dots and to make the necessary flashes with the correct timings.

Thinking about the algorithm for doing this, you can break it into the following steps:

- For each element of the string of dashes and dots (such as .-.-)
  - flash that dot or dash

Using the concept of programming by intention, let's keep the function as simple as that.

The Morse codes are not the same length for all letters, so you need to loop around the string until you encounter the end marker, \0. You also need a counter variable called i that starts at 0 and is incremented as the processing looks at each dot and dash:

```
void flashSequence(char* sequence)
{
    int i = 0;
    while (sequence[i] != '\0')
    {
        flashDotOrDash(sequence[i]);
        i++;
    }
    delay(dotDelay * 3);    // gap between letters
}
```

Again, you delegate the actual job of flashing an individual dot or dash to a new method called **flashDotOrDash**, which actually turns the LED on and off. Finally, when the program has flashed the dots and dashes, it needs to pause for three dots worth of delay. Note the helpful use of a comment.

The *flashDotOrDash* Function

The last function in your chain of functions is the one that actually does the work of turning the LED on and off. As its argument, the function has a single character that is either a dot (.) or a dash (–).

All the function needs to do is turn the LED on and delay for the duration of a dot if it's a dot and three times the duration of a dot if it's a dash, then turn the LED off again. Finally it needs to delay for the period of a dot, to give the gap between flashes.

```
void flashDotOrDash(char dotOrDash)

{
  digitalWrite(ledPin, HIGH);
  if (dotOrDash == '.')
  {
    delay(dotDelay);
  }
  else // must be a -
  {
    delay(dotDelay * 3);
  }
  digitalWrite(ledPin, LOW);
  delay(dotDelay); // gap between flashes
}
```

Putting It All Together

Putting all this together, the full listing is shown in sketch 5-05. Upload it to your Arduino board and try it out. Remember that to use it, you need to open the Serial Monitor and type some text into the area at the top and click Send. You should then see that text being flashed as Morse code.

```
// sketch 5-05
int dotDelay = 200;
int ledPin = 13;

char* letters[] = {
  ".-", "-...", "-.-.", "-..", ".", "..-.", "--.", "....", "..",    // A-I
  ".---", "-.-", ".-..", "--", "-.", "---", ".--.", "--.-", ".-.",  // J-R
  "...", "-", "..-", "...-", ".--", "-..-", "-.--", "--.."          // S-Z
};
```

```
char* numbers[] = {"-----", ".----", "..---", "...--", "....-",
  ".....", "-....", "--...", "---..", "----."};


void setup()
{
  pinMode(ledPin, OUTPUT);
  Serial.begin(9600);
}

void loop()
{
  char ch;
  if (Serial.available() > 0)
  {
    ch = Serial.read();
    if (ch >- 'a' && ch <= 'z')
    {
      flashSequence(letters[ch - 'a']);
    }
    else if (ch >= 'A' && ch <= 'Z')
    {
      flashSequence(letters[ch - 'A']);
    }
    else if (ch >= '0' && ch <= '9')
    {
      flashSequence(numbers[ch - '0']);
    }
    else if (ch == ' ')
    {
     delay(dotDelay * 4);       // gap between words
    }
  }
}

void flashSequence(char* sequence)
{
   int i = 0;
   while (sequence[i] != '\0')
   {
      flashDotOrDash(sequence[i]);
      i++;
   }
   delay(dotDelay * 3);     // gap between letters
}

void flashDotOrDash(char dotOrDash)
{
  digitalWrite(ledPin, HIGH);
  if (dotOrDash == '.')
  {
    delay(dotDelay);
```

```
}
else // must be a -
{
   delay(dotDelay * 3);
}
digitalWrite(ledPin, LOW);
delay(dotDelay); // gap between flashes
}
```

This sketch includes a loop function that is called automatically and repeatedly calls a flashSequence function that you wrote, which itself repeatedly calls a flashDotOrDash function that you wrote, which calls digitalWrite and delay functions that are provided by Arduino!

This is how your sketches should look. Breaking things up into functions makes it much easier to get your code working and makes it easier when you return to it after a period of not using it.

Conclusion

In addition to looking at strings and arrays in this chapter, you have also built this more complex Morse translator that I hope will also reinforce the importance of building your code with functions.

In the next chapter, you learn about input and output, by which we mean input and output of analog and digital signals from the Arduino.

6

Input and Output

The Arduino is about physical computing, and that means attaching electronics to the Arduino board. So you need to understand how to use the various options for your connection pins.

Outputs can be digital, which just means switched between being at 0V or at 5V, or analog, which allows you to set the voltage to any voltage between 0V and 5V—although it's not quite as simple as that, as we shall see.

Likewise, inputs can either be digital (for example, determining whether a button pressed or not) or analog (such as from a light sensor).

In a book that is essentially about software rather than hardware, we are going to try and avoid being dragged into too much discussion of electronics. However, it will help you to understand what is happening in this chapter if you can find yourself a multimeter and a short length of solid core wire.

Digital Outputs

In earlier chapters, you have made use of the LED attached to digital pin 13 of the Arduino board. For example, in Chapter 5, you used it as a Morse code signaler. The Arduino board has a whole load of digital pins available.

Let's experiment with one of the other pins on the Arduino. You will use digital pin 4, and to see what is going on, you will fix some wire to your multimeter leads and attach them to your Arduino. Figure 6-1 shows the arrangement. If your multimeter has crocodile clips, strip the insulation off the ends of some short lengths of solid core wire and attach the clip to one end, fitting the other end into the Arduino socket. If your multimeter does not have crocodile clips then wrap one of the stripped wire ends around the probe.

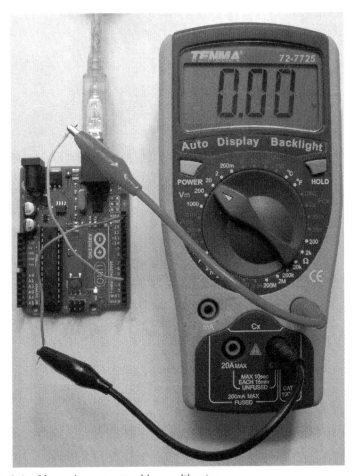

Figure 6-1 *Measuring outputs with a multimeter*

The multimeter needs to be set to its 0–20V direct current (DC) range. The negative lead (black) should be connected to the ground (GND) pin and the positive to D4. The wire is just connected to the probe lead and poked into the socket headers on the Arduino board.

Load sketch 6-01:

```
//sketch 6-01

int outPin = 4;

void setup()
{
  pinMode(outPin, OUTPUT);
  Serial.begin(9600);
  Serial.println("Enter 1 or 0");
}

void loop()
{
  if (Serial.available() > 0)
  {
    char ch = Serial.read();
    if (ch == '1')
    {
      digitalWrite(outPin, HIGH);
    }
    else if (ch == '0')
    {
      digitalWrite(outPin, LOW);
    }
  }
}
```

At the top of the sketch, you can see the command **pinMode**. You should use this command for every pin that you are using in a project so that Arduino can configure the electronics connected to that pin to be either an input or an output, as in the following example:

```
pinMode(outPin, OUTPUT);
```

As you might have guessed, **pinMode** is a built-in function. Its first argument is the pin number in question (an **int**), and the second argument is the mode, which must be either **INPUT** or **OUTPUT**. Note that the mode name must be all uppercase.

This **loop** waits for a command of either **1** or **0** to come from the Serial Monitor on your computer. It it's a **1**, then pin 4 will be turned on; otherwise, it will be turned off.

Upload the sketch to your Arduino and then open the Serial Monitor (shown in Figure 6-2).

So, with the multimeter turned on and plugged into the Arduino, you should be able to see its reading change between 0V and about 5V as you send commands to the board from the Serial Monitor by either pressing **1** and then **Return** or pressing **0** and then **Return**. Figure 6-3 shows the multimeter reading after a **1** has been sent from the Serial Monitor.

If there are not enough pins labeled "D" for your project, you can actually use the pins labeled "A" (for analog) as digital outputs too. To do this, you just have to add *14* to the analog pin number. You could try this out by modifying the first line in sketch 6-01 to use pin 14 and moving your positive multimeter lead to pin A0 on the Arduino.

That is really all there is to digital outputs, so let's move on swiftly to digital inputs.

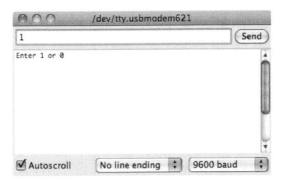

Figure 6-2   *The Serial Monitor*

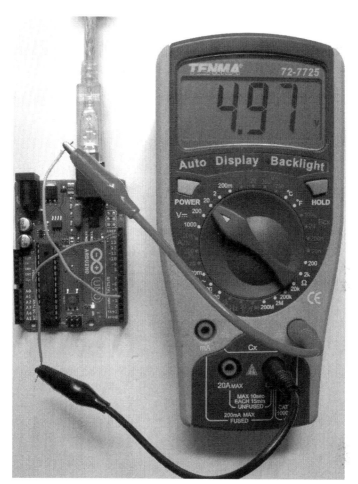

Figure 6-3   *Setting the output to High*

Digital Inputs

The most common use of digital inputs is to detect when a switch has been closed. A digital input can either be on or off. If the voltage at the input is less than 2.5V (halfway to 5V), it will be 0 (off), and if it is above 2.5V, it will be 1 (on).

Disconnect your multimeter and upload the sketch 6-02 onto your Arduino board:

```
//sketch 6-02

int inputPin = 5;

void setup()
{
  pinMode(inputPin, INPUT);
  Serial.begin(9600);
}

void loop()
{
  int reading = digitalRead(inputPin);
  Serial.println(reading);
  delay(1000);
}
```

As with using an output, you need to tell the Arduino in the **setup** function that you are going to use a pin as an input. You get the value of a digital input using the **digitalRead** function. This returns 0 or 1.

Pull-up Resistors

The sketch reads the input pin and writes its value to the Serial Monitor once per second. So upload the sketch and open the Serial Monitor. You should see a value appear once per second. Push one end of your bit of wire into the socket for D5 and pinch the end of the wire between your finger, as shown in Figure 6-4.

Continue pinching for a few seconds and watch the text appear on the Serial Monitor. You should see a mixture of ones and zeros appear in the Serial Monitor. The reason for this is that the inputs to the Arduino board are very sensitive. You are acting as an antenna, picking up electrical interference.

Take the end of the wire that you were holding and push it into the socket for +5V as shown in Figure 6-5. The stream of text in the Serial Monitor should change to ones.

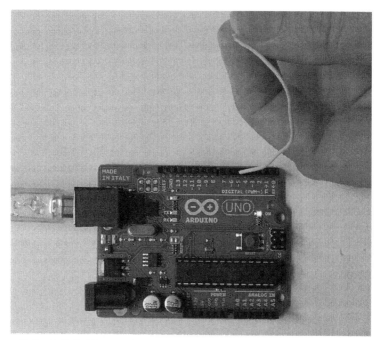

Figure 6-4 *A digital input with a human antenna*

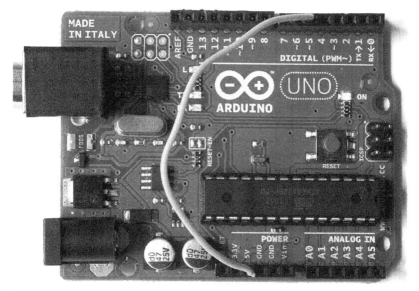

Figure 6-5 *Pin 5 connected to +5V*

Now take the end that was in +5V and put it into one of the GND connections on the Arduino. As you would expect, the Serial Monitor should now display zeros.

A typical use for an input pin is to connect a switch to it. Figure 6-6 shows how you might be expecting to connect your switch.

The problem with this is that if the switch is not closed, then the input pin is not connected to anything. It is said to be floating, and could easily give you a false reading. You need your input to be more predictable, and the way to do this is with what is called a pull-up resistor. Figure 6-7 shows the standard use of a pull-up resistor. It has the effect that if the switch is open, then the resistor pulls up the floating input to 5V. When you press the switch and close the contact, the switch overrides the effect of the resistor, forcing the input to 0V. One side-effect of this is, while the switch is closed, 5V will be across the resistor, causing a current to flow. So, the value of the resistor is selected to be low enough to make it immune from any electrical interference, but at the same time high enough to prevent excessive current drain when the switch is closed.

Figure 6-6 *Connecting a switch to an Arduino board*

Figure 6-7   *Switch with a pull-up resistor*

Internal Pull-up Resistors

Fortunately, the Arduino board has software-configurable pull-up resistors built into the digital pins. By default, they are turned off. So all you need to do to enable the pull-up resistor on pin 5 for sketch 6-02 is to add the following line:

```
digitalWrite(inputPin, HIGH);
```

This line goes in the **setup** function right after you define the pin as an input. It may seem a little strange to do a **digitalWrite** to an input, but this is just the way it works.

Sketch 6-03 is the modified version. Upload it to your Arduino board and test it by acting like an antenna again. You should find that this time the input stays at 1 in the Serial Monitor.

```
//sketch 6-03

int inputPin = 5;

void setup()
{
  pinMode(inputPin, INPUT);
  digitalWrite(inputPin, HIGH);
  Serial.begin(9600);
}
```

```
void loop()
{
    int reading = digitalRead(inputPin);
    Serial.println(reading);
    delay(1000);
}
```

Debouncing

When you press a pushbutton, you would expect that you would just get a single change from 1 (with a pull-up resistor) to 0 as the button is depressed. Figure 6-8 shows what can happen when you press a button. The metal contacts in the button bounce. So a single button press becomes a series of presses that eventually stabilize.

All this happens very quickly; the total time span of the button press on the oscilloscope trace is only 200 milliseconds. This is a very "ropey" old switch. A new tactile, click-type button may not even bounce at all.

Sometimes bouncing does not matter at all. For instance, sketch 6-04 will light the LED while the button is pressed. In reality, you would not

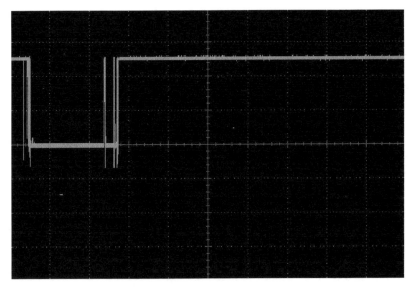

Figure 6-8 *Oscilloscope trace of a button press*

use an Arduino to do this; we are firmly in the realms of theory rather than practice here.

```
//sketch 6-04

int inputPin = 5;
int ledPin = 13;

void setup()
{
  pinMode(ledPin, OUTPUT);
  pinMode(inputPin, INPUT);
  digitalWrite(inputPin, HIGH);
}

void loop()
{
  int switchOpen = digitalRead(inputPin);
  digitalWrite(ledPin, ! switchOpen);
}
```

Looking at the **loop** function of sketch 6-04, the function reads the digital input and assigns its value to a variable **switchOpen**. This is a 0 if the button is pressed and a 1 if it isn't (remember that the pin is pulled up to 1 when the button is not pressed).

When you program **digitalWrite** to turn the LED on or off, you need to reverse this value. You do this using the **!** or **not** operator.

If you upload this sketch and connect your wire between D5 and GND (see Figure 6-9), you should see the LED light. Bouncing may be going on here, but it is probably too fast for you to see and does not matter.

One situation where key bouncing would matter is if you were making your switch toggle the LED on and off. That is, if you press the button, the LED comes on and stays on, and when you press the button again, it turns off. If you had a button that bounced, then whether the LED was on or off would just depend on whether you had an odd or even number of bounces.

Sketch 6-05 just toggles the LED without any attempt at "debouncing." Try it out using your wire as a switch between pin D5 and GND:

```
// sketch 6-05

int inputPin = 5;
```

```
int ledPin = 13;
int ledValue = LOW;

void setup()
{
  pinMode(inputPin, INPUT);
  digitalWrite(inputPin, HIGH);
  pinMode(ledPin, OUTPUT);
}

void loop()
{
  if (digitalRead(inputPin) == LOW)
  {
    ledValue = ! ledValue;
    digitalWrite(ledPin, ledValue);
  }
}
```

You will probably find that sometimes the LED toggles, but other times it appears not to toggle. This is bouncing in action!

A simple way to tackle this problem is simply to add a delay after you detect the first button press, as shown in sketch 6-06:

```
// sketch 6-06
int inputPin = 5;
int ledPin = 13;
int ledValue = LOW;

void setup()
{
  pinMode(inputPin, INPUT);
  digitalWrite(inputPin, HIGH);
  pinMode(ledPin, OUTPUT);
}

void loop()
{
  if (digitalRead(inputPin) == LOW)
  {
    ledValue = ! ledValue;
    digitalWrite(ledPin, ledValue);
    delay(500);
  }
}
```

Figure 6-9 *Using a wire as a switch*

By putting a delay here, nothing else can happen for 500 milliseconds, by which time any bouncing will have subsided. You should find that this makes the toggling much more reliable. An interesting side-effect is that if you hold the button down, the LED just keeps flashing.

If that is all there is to the sketch, then this delay is not a problem. However, if you do more in the **loop**, then using a delay can be a problem; for example, the program would be unable to detect the press of any other button during that 500 milliseconds.

So, this approach is sometimes not good enough and you will need to be a bit more sophisticated. You can write your own advanced debouncing code by hand, but doing so gets complicated and fortunately some fine folks have done all the work for you.

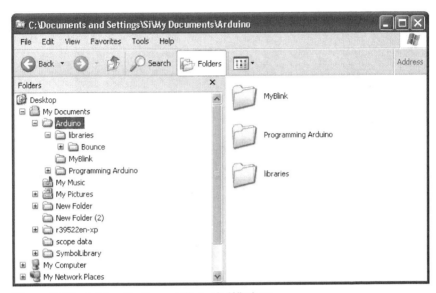

Figure 6-10 *Adding the Bounce library in Windows*

To make use of their work, you must add a library to your Arduino application. To do this, you must download the library itself as a zip file from here: http://www.arduino.cc/playground/Code/Bounce.

After downloading the file, unzip it and place the unzipped folder, called Bounce, into the libraries subfolder in the folder where all your sketches are saved. On Windows, this folder is My Documents\Arduino, and on Mac and Linux, it is Documents/Arduino. If there is no libraries subfolder, then you will need to create one. Figure 6-10 shows the folder structure in Windows after a library has been added.

After adding the library, you need to restart the Arduino application for the changes to take effect. Once you do so, you can use the Bounce library in any sketches that you write.

Sketch 6-07 shows how you can use the Bounce library. Upload it to your board and see how reliable the LED toggling has become.

```
//sketch 6-07
#include <Bounce.h>
```

```
int inputPin = 5;
int ledPin = 13;

int ledValue = LOW;
Bounce bouncer = Bounce(inputPin, 5);

void setup()
{
  pinMode(inputPin, INPUT);
  digitalWrite(inputPin, HIGH);
  pinMode(ledPin, OUTPUT);
}

void loop()
{
    if (bouncer.update() && bouncer.read() == LOW)
    {
        ledValue = ! ledValue;
        digitalWrite(ledPin, ledValue);
    }
}
```

Using the library is pretty straightforward. The first thing that you will notice is this line:

```
#include <Bounce.h>
```

This is necessary to tell the compiler to use the Bounce library.

You then have the following line:

```
Bounce bouncer = Bounce(inputPin, 5);
```

Do not worry about the syntax of this line at the moment; it is actually C++ rather than C syntax, and you will not be meeting C++ until Chapter 11. For now, you will just have to be content to know that this sets up a **bouncer** object for the pin specified, with a debounce period of 5 milliseconds.

From now on, you use that **bouncer** object to find out what the key is doing rather than reading the digital input directly. It has put a kind of debouncing wrapper around your input pin. So, deciding whether a button has been pressed is wrapped up in this line:

```
if (bouncer.update() && bouncer.read() == LOW)
```

The function **update** returns true if something has changed with the **bouncer** object and the second part of the condition checks whether the button went **LOW**.

Analog Outputs

A few of the digital pins—namely digital pins 3, 5, 6, 9, 10, and 11—can provide variable output other than just 5V or nothing. These are the pins on the board with a ~ or "PWM" next to them. PWM stands for Pulse Width Modulation, which refers to the means of controlling the amount of power at the output. It does so by rapidly turning the output on and off.

The pulses are always delivered at the same rate (roughly 500 per second), but the length of the pulses is varied. If you were to use PWM to control the brightness of an LED, then if the pulse were long, your LED would be on all the time. If, however, the pulses are short, then the LED is actually lit only for a small portion of the time. This happens too fast for the observer even to tell that the LED is flickering, and it just appears that the LED is lighter or dimmer.

Before you try using an LED, you can test this out with your multimeter. Set the multimeter up to measure the voltage between GND and pin D3 (see Figure 6-11).

Now upload sketch 6-08 to your board and open the Serial Monitor (see Figure 6-12). Enter the single digit **3** and press Return. You should see your volt meter register about 3V. You can then try any other number between 0 and 5.

```
//sketch 6-08

int outputPin = 3;

void setup()
{
  pinMode(outputPin, OUTPUT);
  Serial.begin(9600);
  Serial.println("Enter Volts 0 to 5");
}
```

```
void loop()
{
  if (Serial.available() > 0)
  {
    char ch = Serial.read();
    int volts = (ch - '0') * 51;
    analogWrite(outputPin, volts);
  }
}
```

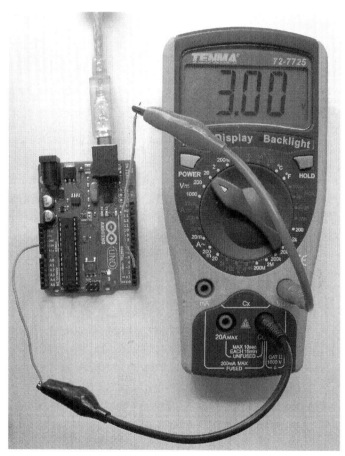

Figure 6-11   *Measuring the analog output*

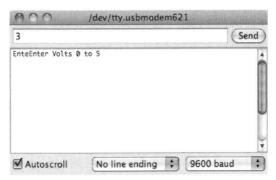

Figure 6-12 *Setting the voltage at an analog output*

The program determines the value of PWM output between 0 and 255 by multiplying the desired voltage (0 to 5) by 51. (Readers may wish to refer to Wikipedia for a fuller description of PWM.)

You can set the value of the output by using the function **analogWrite**, which requires an output value between 0 and 255, where 0 is off and 255 is full power. This is actually a great way to control the brightness of an LED. If you were to try to control the brightness by varying the voltage across the LED, you would find that nothing would happen until you got to about 2V; then the LED would very quickly get quite bright. By controlling the brightness using PWM and varying the average amount of time that the LED is on, you achieve much more linear control of the brightness.

Analog Input

Digital inputs just give you an on/off answer as to what is happening at a particular pin on the Arduino board. Analog inputs, however, give you a value between 0 and 1023 depending on the voltage at the analog input pin.

The program reads the analog input using the **analogRead** function. Sketch 6-09 displays the reading and actual voltage at the analog pin A0 in the Serial Monitor every half second, so open the Serial Monitor and watch the readings appear.

```
//sketch 6-09

int analogPin = 0;

void setup()
{
  Serial.begin(9600);
}

void loop()
{
  int reading = analogRead(analogPin);
  float voltage = reading / 204.6;
  Serial.print("Reading=");
  Serial.print(reading);
  Serial.print("\t\tVolts=");
  Serial.println(voltage);
  delay(500);
}
```

When you run this sketch, you will notice that the readings change quite a bit. As with the digital inputs, this is because the input is floating.

Take one end of the wire and put it into a GND socket so that A0 is connected to GND. Your readings should now stay at 0. Move the end of the lead that was in GND and put it into 5V and you should get a reading of around 1023, which is the maximum reading. So, if you were to connect A0 to the 3.3V socket on the Arduino board, the Arduino volt meter should tell you that you have about 3.3V.

Conclusion

This concludes our chapter on the basics of getting signals into and out of the Arduino. In the next chapter, we will look at some of the features provided in the standard Arduino library.

7
The Standard Arduino Library

This library is where all the goodies live. You can only get so far with the core C language; what you really need is a big collection of functions that you can use in your sketches.

You have already met a fair few of these, such as **pinMode**, **digitalWrite**, and **analogWrite**. But actually, there are many more. There are functions that you can use for doing math, making random numbers, manipulating bits, detecting pulses on an input pin, and using something called interrupts.

The Arduino language is based on an earlier library called Wiring and it complements another library called Processing. The Processing library is very similar to Wiring, but it is based on the Java language rather than C and is used on your computer to link to Android over USB. In fact, the Arduino application that you run on your computer is based on Processing. If you find yourself wanting to write some fancy interface on your computer to talk to an Arduino, then take a look at Processing (www.processing.org).

Random Numbers

Despite the experience of anyone using a PC, computers are in actual fact very predictable. Occasionally it is useful to be able to deliberately make your Arduino unpredictable. For example, you might want to make a robot take a "random" path around a room, heading for a random amount of time in one direction, turning a random number of degrees, and then setting off again. Or, you might be contemplating making an Arduino-based die that gives you a random number between one and six. The Arduino standard library provides you with a feature to do just this. It is the function called **random**. **random** returns an **int** and it can take either one argument or two. If it just takes one argument, then it will return a random number between zero and the argument minus one.

The two argument version produces a random number between the first argument (inclusive) and the second argument minus one. Thus **random(1, 10)** produces a random number between one and nine.

Sketch 7-01 pumps out numbers between one and six to the Serial Monitor.

```
//sketch 7-01

void setup()
{
  Serial.begin(9600);
}

void loop()
{
  int number = random(1, 7);
  Serial.println(number);
  delay(500);
}
```

If you upload this sketch to your Arduino and open the Serial Monitor, you will see something like Figure 7-1.

If you run this a few times you will probably be surprised to see that, every time you run the sketch, you get the same series of 'random' numbers.

The output is not really random; the numbers are called *pseudo-random* numbers because they have a random distribution. That is, if you ran this sketch and collected a million numbers, you would get pretty much the same number of ones, twos, threes, and so on. The numbers are not random in the sense of being unpredictable. In fact, it is so against the

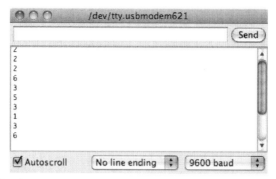

Figure 7-1 *Random numbers*

workings of a microcontroller to be random that it just plain can't do it without some intervention from the real world.

You can provide this intervention to make your sequence of numbers less predictable by *seeding* the random number generator. This basically just gives it a starting point for the sequence. But, if you think about it, you cannot just use **random** to seed the random number generator. A commonly used trick is to use the fact that (as discussed in the previous chapter) an analog input will float. So you can use the value read from an analog input to seed the random number generator.

The function that does this is called **randomSeed**. Sketch 7-02 shows how you can add a bit more randomness to your random number generator.

```
//sketch 7-02

void setup()
{
  Serial.begin(9600);
  randomSeed(analogRead(0));
}

void loop()
{
  int number = random(1, 7);
  Serial.println(number);
  delay(500);
}
```

Try pressing the Reset button a few times. You should now see that your random sequence is different every time. This type of random number generation could not be used for any kind of lottery. For much better random number generation, you would need hardware random number generation, which is sometimes based on random events, such as cosmic ray events.

Math Functions

On rare occasions, you will need to do a lot of math on an Arduino, over and above the odd bit of arithmetic. But, should you need to, there is a big library of math functions available to you. The most useful of these functions are summarized in the following table:

Function	Description	Example
abs	Returns the unsigned value of its argument.	abs(12) returns 12 abs(-12) returns 12
constrain	Constrains a number to stop it from exceeding a range. The first argument is the number to constrain, the second is the start of the range, and the third is the end of the allowed range of numbers.	constrain(8, 1, 10) returns 8 constrain(11, 1, 10) returns 10 constrain(0, 1, 10) returns 1
map	Maps a number in one range into another range. The first argument is the number to map, the second and third are the "from" range (or source range), and the last two are the "to" range (or destination range). The function is useful for remapping analog input values.	map(x, 0, 1023, 0, 5000)
max	Returns the larger of its two arguments.	max(10, 11) returns 11
min	Returns the smaller of its two arguments.	min(10, 11) returns 10
pow	Returns the first argument raised to the power of the second argument.	pow(2, 5) returns 32
sqrt	Returns the square root of a number.	sqrt(16) returns 4
sin, cos, tan	Perform trigonometric functions. They are not often used.	
log	Calculates the temperature from a logarithmic thermistor (for example).	

Bit Manipulation

A bit is a single digit of binary information, that is, either 0 or 1. The word *bit* is a contraction of *binary digit*. Most of the time, you use **int** variables that actually comprise 16 bits This is a bit wasteful if you only need to store a simple true/false value (1 or 0). Actually, unless you are running short of memory, being wasteful is less of a problem than creating difficult-to-understand code, but sometimes it is useful to be able to pack your data tightly.

Each bit in the **int** can be thought of as having a decimal value, and you can find the decimal value of the **int** by adding up the values of all the bits that are a 1. So in Figure 7-2, the decimal value of the **int** would be 38. Actually, it gets more complicated to deal with negative numbers, but that only happens when the leftmost bit becomes a 1.

When you are thinking about individual bits, decimal values do not really work very well. It is very difficult to visualize which bits are set in a decimal number such as 123. For that reason, programmers often use something called *hexadecimal*, or, more commonly, just *hex*. Hex is number base 16. So instead of having digits 0 to 9, you have six extra digits, A to F. This means that each hex digit represents four bits. The following table shows the relationship among decimal, hex, and binary with the numbers 0 to 15:

Decimal	Hex	Binary (Four Digit)
0	0	0000
1	1	0001
2	2	0010
3	3	0011
4	4	0100
5	5	0101
6	6	0110
7	7	0111
8	8	1000
9	9	1001
10	A	1010

Decimal	Hex	Binary (Four Digit)
11	B	1011
12	C	1100
13	D	1101
14	E	1110
15	F	1111

So, in hex, any **int** can be represented as a four-digit hex number. Thus, the binary number 10001100 would in hex be 8C. The C language has a special syntax for using hex numbers. You can assign a hex value to an **int** as follows:

```
int x = 0x8C;
```

The Arduino standard library provides some functions that let you manipulate the 16 bits within an **int** individually. The function **bitRead** returns the value of a particular bit in an **int**; so, for the following example would assign the value 0 to the variable called **bit**:

```
int x = 0x8C; // 10001100
int bit = bitRead(x, 0);
```

In the second argument, the bit position starts at 0 and goes up to 15. It starts with the least significant bit. So the rightmost bit is bit 0, the next bit to the left is bit 1, and so on.

As you would expect, the counterpart to **bitRead** is **bitWrite**, which takes three arguments. The first is the number to manipulate, the second is the bit position, and the third is the bit value. The following example changes the value of the **int** from 2 to 3 (in decimal or hex):

```
int x = 2; // 0010
bitWrite(x, 0, 1);
```

16384 8192 4096 2048 1024 512 256   128 64   32   16   8   4   2   1

| 0 | 0 | 0 | 0 | 0 | 0 | 0 | 0 | 0 | 0 | 1 | 0 | 0 | 1 | 1 | 0 |

32 + 4 + 2 = 38

Figure 7-2 *An* int

Advanced I/O

There are some useful little functions that you can use to make your life easier when performing various input/output tasks.

Generating Tones

The **tone** function allows you to generate a square-wave signal (see Figure 7-3) on one of the digital output pins. The most common reason to do this is to generate an audible tone using a loudspeaker or buzzer.

The function takes either two or three arguments. The first argument is always the pin number on which the tone is to be generated, the second argument is the frequency of the tone in hertz (Hz), and the optional final argument is the duration of the tone. If no duration is specified, then the tone will continue playing indefinitely, as is the case in sketch 7-03. This is

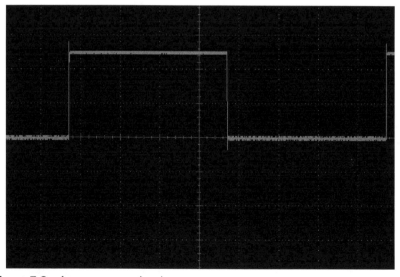

Figure 7-3 *A square-wave signal*

why we have put the **tone** function call in **setup** rather than in the **loop** function.

```
//sketch 7-03

void setup()
{
  tone(4, 500);
}

void loop() {}
```

To stop a tone that is playing, you use the function **noTone**. This function has just one argument, which is the pin on which the tone is playing.

Feeding Shift Registers

Sometimes the Arduino Uno just doesn't have enough pins. When driving a large number of LEDs, for example, a common technique is to use a shift register chip. This chip reads data one bit at a time, and then when it has enough, it latches all those bits onto a set of outputs (one per bit).

To help you use this technique, there is a handy function called **shift-Out**. This function takes four arguments:

- The number of the pin on which the bit to be sent will appear.

- The number of the pin to be used as a clock pin. This toggles every time a bit is sent.

- A flag to determine whether the bits will be sent starting with the least significant bit or the most significant. This should be one of the constants **MSBFIRST** or **LSBFIRST**.

- The byte of data to be sent.

Interrupts

One of the things that tend to frustrate programmers used to "programming in the large" is that the Arduino can do only one thing at a time. If you like to have lots of threads of execution all running at the same time in your programs, then you are out of luck. Although a few people have developed projects that can execute multiple threads in this way, generally this capability is unnecessary for the type of uses that an Arduino is normally put to. The closest an Arduino gets to such execution is the use of interrupts.

Two of the pins on the Arduino (D2 and D3) can have interrupts attached to them. That is, these pins act as inputs that, if the pins receive a signal in a specified way, the Arduino's processor will suspend whatever it was doing and run a function attached to that interrupt.

Sketch 7-04 blinks an LED, but then changes the blink period when an interrupt is received. You can simulate an interrupt by connecting your wire between pin D2 and GND and using the internal pull-up resistor to keep the interrupt high most of the time.

```
//sketch 7-04

int interruptPin = 2;
int ledPin = 13;
int period = 500;

void setup()
{
  pinMode(ledPin, OUTPUT);
  pinMode(interruptPin, INPUT);
  digitalWrite(interruptPin, HIGH); // pull-up
  attachInterrupt(0, goFast, FALLING);
}

void loop()
{
  digitalWrite(ledPin, HIGH);
  delay(period);
  digitalWrite(ledPin, LOW);
  delay(period);
}

void goFast()
{
  period = 100;
}
```

The following is the key line in the **setup** function of this sketch:

```
attachInterrupt(0, goFast, FALLING);
```

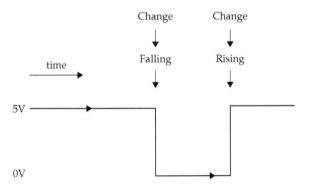

Figure 7-4   *Types of interrupt signals*

The first argument specifies which of the two interrupts you want to use. Rather confusingly, a 0 here means you are using pin 2, while a 1 means you are using pin 3.

The next argument is the name of the function that is to be called when there is an interrupt, and the final argument is a constant that will be one of **CHANGE, RISING,** or **FALLING.** Figure 7-4 summarizes these options.

If the interrupt mode is CHANGE, then either a RISING from 0 to 1 or a FALLING from 1 to 0 will both trigger an interrupt.

You can disable interrupts using the function **noInterrupts**. This stops all interrupts from both interrupt channels. You can resume using interrupts again by calling the function **interrupts**.

Conclusion

In this chapter, you have looked at some of the handy features that the Arduino standard library provides. These features will save you some programming effort, and if there is one thing that a good programmer likes, it is being able to use high-quality work done by other people.

In the next chapter, we will extend what we learned about data structures in Chapter 5 and look at how you go about remembering data on the Arduino after the power goes off.

8

Data Storage

When you give values to variables, the Arduino board will remember those values only as long as the power is on. The moment that you turn the power off or reset the board, all that data is lost.

In this chapter, we look at some ways to hang on to that data.

Constants

If the data that you want to store does not change, then you can just set the data up each time that the Arduino starts. An example of this approach is the case in the letters array in your Morse code translator of Chapter 5 (sketch 5-05).

You used the following code to define a variable of the correct size and fill it with the data that you needed:

```
char* letters[] = {
    ".-", "-...", "-.-.", "-..", ".",
    "..-.", "--.", "....", "..",       // A-I
    ".---", "-.-", ".-..", "--", "-.",
    "---", ".--.", "--.-", ".-.",      // J-R
    "...", "-", "..-", "...-", ".--",
    "-..-", "-.--", "--.."             // S-Z
};
```

You may remember that you did the calculation and decided that you had plenty of your meager 2K to spare. However, if memory was a bit tight, it would be far better to be able to store this data in the 32K of flash

115

memory used to store programs, rather than the 2K of RAM. There is a means of doing this. It is a directive called **PROGMEN**; it lives in a library and is a bit awkward to use.

The PROGMEM Directive

To store your data in flash memory, you have to include the **PROGMEM** library as follows:

```
#include <avr/pgmspace.h>
```

The purpose of this command is to tell the compiler to use the **pgmspace** library for this sketch. In this case, a library is a set of functions that someone else has written and that you can use in your sketches without having to understand all the details of how those functions work.

Because you are using this library, the **PROGMEM** keyword and the **pgm_read_word** function are available. You will use both in the sketches that follow.

This library is included as part of the Arduino software and is an officially supported Arduino library. A good collection of such official libraries is available, and many unofficial libraries, developed by people like you and made for others to use, are also available on the Internet. Such unofficial libraries must be installed into your Arduino environment. You will learn more about these libraries, as well as how to write your own libraries, in Chapter 11.

When using **PROGMEM**, you have to make sure that you use special **PROGMEM**-friendly data types. Unfortunately, that does *not* include an array of **char** arrays. You actually have to define a variable for each string using a **PROGMEM** string type and then put them all in a **PROGMEM** array type, like this:

```
PROGMEM prog_char sA[]   = ".-";
PROGMEM prog_char sB[]   = "-...";
// and so on for all letters
PROGMEM const char* letters[] =
{
  sA, sB, sC, sD, sE, sF, sG, sH, sI, sJ, sK, sL, sM,
  sN, sO, sP, sQ, sR, sS, sT, sU, sV, sW, sX, sY, sZ
};
```

I have not listed sketch 8-01 here, as it is a little lengthy, but you may wish to load it and verify that it works the same way as the RAM-based version.

In addition to creating the data in a special way, you also have to read the data back a special way. Your code to get the code string for a Morse letter from the array has to be modified to look like this:

```
strcpy_P(buffer, (char*)pgm_read_word(&(letters[ch - 'a'])));
```

This uses a **buffer** variable into which the **PROGMEM** string is copied, so that it can be used as a regular **char** array. This needs to be defined as a global variable as follows:

```
char buffer[6];
```

This approach works only if the data is constant—that is, you are not going to change it while the sketch is running. In the next section, you will learn about using the EEPROM memory that is intended for storing persistent data that can be changed.

EEPROM

The ATMega328 at the heart of an Arduino Uno has a kilobyte of electrically erasable read-only memory (EEPROM). EEPROM is designed to remember its contents for many years. Despite its name, it is not really read-only. You can write to it.

The Arduino commands for reading and writing to EEPROM are just as awkward to use as the ones for using **PROGMEM**. You have to read and write to and from EEPROM one byte at a time.

The example of sketch 8-02 allows you to enter a single-digit letter code from the Serial Monitor. The sketch then remembers the digit and repeatedly writes it out on the Serial Monitor.

```
// sketch 8-02
#include <EEPROM.h>

int addr = 0;
char ch;
```

```
void setup()
{
  Serial.begin(9600);
  ch = EEPROM.read(addr);
}

void loop()
{
  if (Serial.available() > 0)
  {
    ch = Serial.read();
    EEPROM.write(0, ch);
    Serial.println(ch);
  }
  Serial.println(ch);
  delay(1000);
}
```

To try this sketch, open the Serial Monitor and enter a new character. Then unplug the Arduino and plug it back in. When you reopen the Serial Monitor, you will see that the letter has been remembered.

The function **EEPROM.write** takes two arguments. The first is the address, which is the memory location in EEPROM and should be between 0 and 1023. The second argument is the data to write at that location. This must be a single byte. A character is represented as eight bits, so this is fine, but you cannot directly store a 16-bit **int**.

Storing an *int* in EEPROM

To store a two-byte **int** in locations 0 and 1 of the EEPROM, you would have to do this:

```
int x = 1234;
EEPROM.write(0, highByte(x));
EEPROM.write(1, lowByte(x));
```

The functions **highByte** and **lowByte** are useful for separating an **int** into two bytes. Figure 8-1 shows how this **int** is actually stored in the EEPROM.

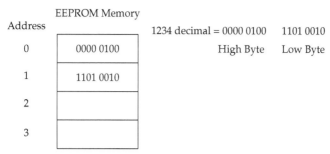

Figure 8-1 *Storing a 16-bit integer in EEPROM*

To read the **int** back out of EEPROM, you need to read the two bytes from the EEPROM and reconstruct the **int**, as follows:

```
byte high = EEPROM.read(0);
byte low = EEPROM.read(1);
int x = (high << 8) + low;
```

The << operator is a bit shift operator that moves the eight high bytes to the top of the **int** and then adds in the low byte.

Storing a *float* in EEPROM (Unions)

Storing a **float** in EEPROM is a little more tricky. To do this, you can use a feature of C called **union**s. These data structures are interesting in that they can be thought of as a way to make the same area of memory accessible to more than one variable. What is more, these variables are allowed to be of different types as long as they are the same size in bytes.

The following **union** definition allows both a **float** and an **int** to refer to the same two bytes of memory:

```
union data
{
  float f;
  int i;
} convert;
```

You can then put a **float** into the **union** as follows:

```
float f = 1.23;
convert.f = f;
```

Then you can separate an integer into its two bytes for storage in EEPROM as follows:

```
EEPROM.write(0, highByte(convert.i));
EEPROM.write(1, lowByte(convert.i));
```

Reading the **float** back out again requires you to do the reverse. First you assemble the two bytes into a single **int**, then you put the **int** into the **union** and pull it out again as a **float**.

```
byte high = EEPROM.read(0);
byte low = EEPROM.read(1);
convert.i = (high << 8) + low;
float f = convert.f;
```

Storing a String in EEPROM

Writing and reading character strings into the EEPROM are pretty straightforward; you just have to write each character at a time, as in the following example:

```
char *test = "Hello";
int i = 0;
while (test[i] != '\0')
{
  EEPROM.write(i, test[i]);
  i++;
}
```

To read the string back into a character array, you can do something like this:

```
char test[10];
int i = 0;
char ch;
ch = EEPROM.read(i);
while (ch != '\0' && i < 10)
{
  test[i] = ch;
  ch = EEPROM.read(i);
  i++;
}
```

Clearing the Contents of EEPROM

When writing to EEPROM, remember that even uploading a new sketch will not clear the EEPROM, so you may have leftover values in there from a previous project. Sketch 8-03 resets all the contents of EEPROM to zeros:

```
// sketch 8-03
#include <EEPROM.h>

void setup()
{
  Serial.begin(9600);
  Serial.println("Clearing EEPROM")
  for (int i = 0; i < 1023; i++)
  {
    EEPROM.write(i, 0);
  }
  Serial.println("EEPROM Cleared");
}

void loop()
{
}
```

Also be aware that you can write to an EEPROM location only about 100,000 times before it will become unreliable. So only write a value back to EEPROM when you really need to. EEPROM is also quite slow, taking about 3 milliseconds to write a byte.

Compression

When saving data to EEPROM or when using **PROGMEM**, you will sometimes find that you have more to save than you have room to save it. When this happens, it is worth finding the most efficient way of representing the data.

Range Compression

You may have a value for which on the face of it you need an **int** or a **float** that are both 16-bit. For example, to represent a temperature in degrees

Celsius, you might use a **float** value such as 20.25. When you are storing that into EEPROM, life would be so much easier if you could fit it into a single byte, and you could store twice as much as if you used a **float**. One way that you can do this is to change the data before you store it. Remember that a byte will allow you to store a positive number between 0 and 255. So if you only cared about the temperature to the nearest degree Celsius, then you could simply convert the **float** to an **int** and discard the part after the decimal point. The following example shows how to do this:

```
int tempInt = (int)tempFloat;
```

The variable **tempFloat** contains the floating point value. The **(int)** command is called a *type cast* and is used to convert a variable from one type to another compatible type. In this case, the type cast converts the **float** of (for example) 20.25 to an **int** that will simply truncate the number to 20.

If you know that the highest temperature that you care about is 60 degrees Celsius and that the lowest is 0 degrees Celsius, then you could multiply every temperature by 4 before converting it to a byte and saving it. Then when you read the data back from EEPROM, you can divide by 4 to get a value that has a precision of 0.25 of a degree.

The following code example (sketch 8-04) saves such a temperature into EEPROM, then reads it back and displays it in the Serial Monitor as proof:

```
//sketch 8-04

#include <EEPROM.h>

void setup()
{
  float tempFloat = 20.75;
  byte tempByte = (int)(tempFloat * 4);
  EEPROM.write(0, tempByte);

  byte tempByte2 = EEPROM.read(0);
  float temp2 = (float)(tempByte2) / 4;
  Serial.begin(9600);
  Serial.println("\n\n\n");
  Serial.println(temp2);
}

void loop(){}
```

There are other means of compressing data. For instance, if you are taking readings that change slowly—again, changes in temperature are a good example of this—then you can record the first temperature at full resolution and then just record the changes in temperature from the previous reading. This change will generally be small and occupy fewer bytes.

Conclusion

You now know a little about how to make your data hang around after the power has gone off. In the next chapter, you will look at LCD displays.

9

LCD Displays

In this chapter, you look at how to write software to control LCD displays. Figure 9-1 shows the kind of LCD display used.

This is a book about software, not hardware, but in this chapter, we will have to explain a little about how the electronics of these displays work so that you understand how to drive them.

The LCD module that we use is a prebuilt Arduino shield that can just be plugged on top of an Arduino board. In addition to its display, it also has some buttons. There are a number of different shields, but nearly all of them use the same LCD controller chip (the HD44780), so look for a shield that uses this controller chip.

I used the DFRobot LCD Keypad Shield for Arduino. This module supplied by DFRobot (www.robotshop.com) is inexpensive and provides an LCD display that is 16 characters by 2 rows and also has six pushbuttons.

The shield comes assembled, so no soldering is required; you just plug it on top of your Arduino board (see Figure 9-2).

The LCD shield uses seven of the Arduino pins to control the LCD display and one analog pin for the buttons. So we cannot use these Arduino pins for any other purpose.

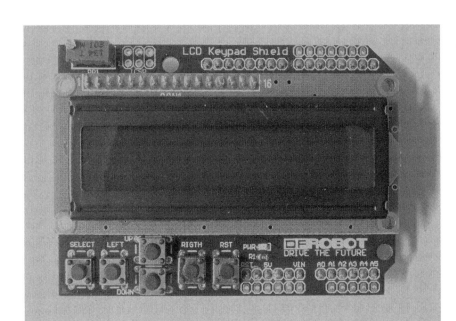

Figure 9-1 *An Alphanumeric LCD shield*

Figure 9-2 *LCD shield attached to an Arduino board*

A USB Message Board

For a simple example of a simple use of the display, we are going to make a USB message board. This will display messages sent from the Serial Monitor.

The Arduino IDE comes with an LCD library. This greatly simplifies the process of using an LCD display. The library gives you useful functions that you can call:

- **clear** clears the display of any text.

- **setCursor** sets the position in row and column where the next thing that you print will appear.

- **print** writes a string at that position.

This example is listed in sketch 9-01:

```
// sketch 9-01 USB Message Board

#include <LiquidCrystal.h>

// lcd(RS, E, D4, D5, D6, D7)
LiquidCrystal lcd(8, 9, 4, 5, 6, 7);
int numRows = 2;
int numCols = 16;

void setup()
{
  Serial.begin(9600);
  lcd.begin(numRows, numCols);
  lcd.clear();
  lcd.setCursor(0,0);
  lcd.print("Arduino");
  lcd.setCursor(0,1);
  lcd.print("Rules");
}

void loop()
{
  if (Serial.available() > 0)
  {
```

```
   char ch = Serial.read();
   if (ch == '#')
   {
      lcd.clear();
   }
   else if (ch == '/')
   {
      // new line
      lcd.setCursor(0, 1);
   }
   else
   {
      lcd.write(ch);
   }
 }
}
```

As with all Arduino libraries, you have to start by including the library to make the compiler aware of it.

The next line defines which Arduino pins are used by the shield and for what purpose. If you are using a different shield, then you may well find that the pin allocations are different, so check in the documentation for the shield.

In this case, the six pins used to control the display are D4, D5, D6, D7, D8, and D9. The purpose of each of these pins is described in Table 9-1.

Parameter to LCD()	Arduino Pin	Purpose
RS	8	Register Select; this is set to 1 or 0 depending on whether the Arduino is sending data for characters or an instruction. An instruction might make the cursor flash, for example.
E	9	Enable; this gets toggled to tell the LCD controller chip that the data on the following four pins is ready to be read.
Data 4	4	These four pins are used to transfer data. The LCD controller chip used by the shield can use eight-bit or four-bit data. This shield uses four bits, in which case the bits 4–7 are used rather than 0–7.
Data 5	5	
Data 6	6	
Data 7	7	

Table 9-1 *LCD shield pin assignments*

The **setup** function is straightforward. You start serial communications so that the Serial Monitor can send commands and initialize the LCD library with the dimensions of the display being used. You also display the message "Arduino Rules" on two lines by setting the cursor to top-left, printing "Arduino," then moving the cursor to the start of the second row and printing "Rules."

Most of the action takes place in the **loop** function, which checks for any incoming characters from the Serial Monitor. The sketch deals with characters one at a time.

Apart from ordinary characters that the sketch will simply display, there are also a couple of special characters. If the character is a **#**, then the sketch clears the whole display, and if the character is a **/**, the sketch moves to the second line. Otherwise, the sketch simply displays the character at the current cursor position using **write**. The function **write** is like **print**, but it prints only a single character rather than a string of characters.

Using the Display

Try out sketch 9-01 by uploading it to the board and then attaching the shield. Note that you should always unplug the Arduino board so that it is off before you plug in a shield.

Open up the Serial Monitor and try typing in the text shown in Figure 9-3.

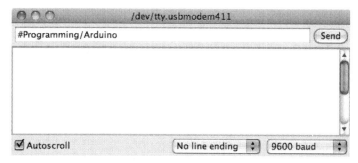

Figure 9-3 *Sending commands to the display*

Other LCD Library Functions

In addition to the functions that you have used in this example, there are a number of other functions that you can use:

- **home** is the same as **setCursor(0,0)**: it moves the cursor to top-left.

- **cursor** displays a cursor.

- **noCursor** specifies not to display a cursor.

- **blink** makes the cursor blink.

- **noBlink** stops the cursor from blinking.

- **noDisplay** turns off the display without removing the content.

- **display** turns the display back on after **noDisplay**.

- **scrollDisplayLeft** moves all the text on the display one character position to the left.

- **scrollDisplayRight** moves all the text on the display one character position to the right.

- **autoscroll** activates a mode in which, as new characters are added at the cursor, the existing text is pushed in the direction determined by the functions **leftToRight** and **rightToLeft.**

- **noAutoscroll** turns **autoscroll** mode off.

Conclusion

You can see that programming shields is not hard, particularly when there is a library that can do a lot of the work.

In the next chapter, you will use an Ethernet shield that will allow you to connect the Arduino to the Internet.

10

Arduino Ethernet Programming

In this chapter, you will use an Ethernet shield to enable your Arduino to work over your home network (see Figure 10-1).

Figure 10-1 *Arduino with Ethernet*

Ethernet Shields

When buying an Ethernet shield, you need to take a little care, as you need to use an "official" shield based on the Wiznet chipset and *not* one of the cheaper but more difficult to use unofficial boards based on the ENC28J60 Ethernet controller chip.

The Ethernet shields are quite power hungry, so you will also need a 9V or 12V power supply rated at 1A or more. This supply will be attached to the Arduino power socket.

Communicating with Web Servers

Before looking at how the Arduino deals with communication between a browser and the web server that it uses, you need some understanding of the HyperText Transfer Protocol (HTTP) and the HyperText Markup Language (HTML).

HTTP

The HyperText Transport Protocol is the method by which web browsers communicate with a web server.

When you go to a page using a web browser, the browser sends a request to the server hosting that page, saying what it wants. What the browser asks for may be simply the contents of a page in HTML. The web server is always listening for such requests, and when it receives one, it processes it. In this simple case, processing the request just means sending back HTML that you have specified in the Arduino sketch.

HTML

The HyperText Markup Language is a way of adding formatting to ordinary text so that it looks good when the browser displays it. For example, the following code is HTML that displays on a browser page as shown in Figure 10-2:

```
<html>
<body>
```

```
<h1>Programming Arduino</h1>
<p>A book about programming Arduino</p>
</body>
</html>
```

The HTML contains tags. Tags have a start and an end and usually contain other tags. The start of a tag has a < and then the tag name, and then a >; for example, **<html>**. The end of a tag is similar except that it has a / after the <. In the preceding example, the outermost tag is **<html>** that contains a tab called **<body>**. All web pages should start with such tags, and you can see the corresponding ends for those tags at the end of the file. Note that you have to put the end tags in the right order, so the **body** tag must be closed before the **html** tag.

Now we get to the interesting bit in the middle, the **h1** and **p** tags. These are the parts of the example that are actually displayed.

The **h1** tag indicates a level 1 header. This has the effect of displaying the text that it contains in a large bold font. The **p** tag is a paragraph tag, and so all the text contained within it is displayed as a paragraph.

This really just scratches the surface of HTML. Many books and Internet resources are available for learning about HTML.

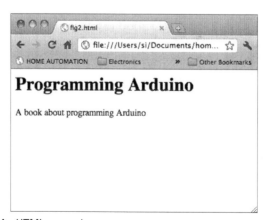

Figure 10-2   *An HTML example*

Arduino as a Web Server

The first example sketch simply uses the Arduino and Ethernet shield to make a small web server. It's definitely not a Google server farm, but it will allow you to send a web request to your Arduino and view the results in a browser on your computer.

Before uploading the sketch 10–01, there are a couple of changes that you need to make. If you look at the top of the sketch, you will see the following lines:

```
byte mac[] = { 0xDE, 0xAD, 0xBE, 0xEF, 0xFE, 0xED };
byte ip[] = { 192, 168, 1, 30 };
```

The first of these, the **mac** address, must be unique among all the devices connected to your network. The second one is the IP address. Whereas most devices that you connect to your home network will have IP addresses assigned to them automatically by a protocol called Dynamic Host Configuration Protocol (DHCP), this is not true for the Ethernet shield. For this device, you have to define an IP address manually. This address cannot be any four numbers; they must be numbers that qualify as being internal IP addresses and fit in the range of IP addresses expected by your home router. Typically, the first three numbers will be something like 10.0.1.x or 192.168.1.x, where x is a number between 0 and 255. Some of these IP addresses will be in use by other devices on your network. To find an unused but valid IP address, connect to the administration page for your home router and look for an option that says "DHCP." You should find a list of devices and their IP addresses, similar to that shown in Figure 10-3. Select a final number to use in you IP address. In this case, 192.168.1.30 looked like a good bet, and indeed it worked fine.

Attach the Arduino to your computer using the USB lead and upload the sketch. You can now disconnect the USB lead and attach the power supply to the Arduino and the Ethernet lead.

Open a connection on your computer's browser to the IP address that you assigned for the Ethernet shield. Something very much like Figure 10-4 should appear.

Figure 10-3   *Finding an unused IP address*

Figure 10-4   *A Simple Arduino server example*

The listing for sketch 10-01 is as follows:

```
// sketch 10-01 Simple Server Example

#include <SPI.h>
#include <Ethernet.h>


// MAC address just has to be unique. This should work:
byte mac[] = { 0xDE, 0xAD, 0xBE, 0xEF, 0xFE, 0xED };
// The IP address will be dependent on your local network:
byte ip[] = { 192, 168, 1, 30 };

Server server(80);

void setup()
{
  Ethernet.begin(mac, ip);
  server.begin();
  Serial.begin(9600);
}

void loop()
{
  // listen for incoming clients
  Client client = server.available();
  if (client)
  {
    while (client.connected())
    {
      // send a standard http response header
      client.println("HTTP/1.1 200 OK");
      client.println("Content-Type: text/html");
      client.println();

      // send the body
      client.println("<html><body>");
      client.println("<h1>Arduino Server</h1>");
      client.print("<p>A0=");
      client.print(analogRead(0));
      client.println("</p>");
      client.print("<p>millis=");
      client.print(millis());
      client.println("</p>");
      client.println("</body></html>");
      client.stop();
    }
    delay(1);
  }
}
```

As with the LCD library discussed in Chapter 9, a standard Arduino library takes care of interfacing with the Ethernet shield.

The **setup** function initializes the Ethernet library using the **mac** and IP addresses that you set earlier.

The **loop** function is responsible for servicing any requests that come to the web server from a browser. If a request is waiting for a response, then calling **server.available** will return a client. A client is an object; you will learn a bit more about what this means in Chapter 11. But for now, all that you need to know is that whether a client exists (tested by the first **if** statement); then you can then determine whether it is connected to the web server by calling **client.connected**.

The next three lines of code print out a return header. This just tells the browser what type of content to display. In this case, the browser is to display HTML content.

Once the header has been written, all that remains is to write the remaining HTML back to the browser. This must include the usual **<html>** and **<body>** tags, and also includes a **<h1>** header tag and two **<p>** tags that will display the value on the analog input A0 and the value returned by the **millis** function; that value is the number of milliseconds since the Arduino was last reset.

Finally, **client.stop** tells the browser that the message is complete. The browser then displays the page.

Setting Arduino Pins over the Network

This second example of using an Ethernet shield allows you to turn the Arduino pins D3 to D7 on and off using a web form.

Unlike the simple server example, you are going to have to find a way to pass the pin settings to the Arduino.

The method for doing this is called *posting data* and is part of the HTTP standard. For this method to work, you have to build the posting mechanism into the HTML so that the Arduino returns HTML to the browser, which renders a form. This form (shown in Figure 10-5) has a selection of

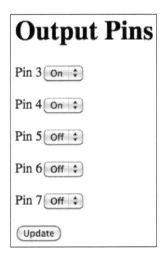

Figure 10-5 *The message sending form*

On and Off for each pin and an Update button that will send the pin settings to the Arduino.

When the Update button is pressed, a second request is sent to the Arduino. This will be just like the first request, except that the request will contain request parameters that will contain the values of the pins.

A request parameter is similar in concept to a function parameter. A function parameter enables you to get information to a function, such as the number of times to blink, and a request parameter enables you to pass data to the Arduino that is going to handle the web request. When the Arduino receives the web request, it can extract the pin settings from the request parameter and change the actual pins.

The code for the second example sketch follows:

```
// sketch 10-02 Internet Pins

#include <SPI.h>
#include <Ethernet.h>

// MAC address just has to be unique. This should work:
byte mac[] = { 0xDE, 0xAD, 0xBE, 0xEF, 0xFE, 0xED };
// The IP address will be dependent on your local network:
byte ip[] = { 192, 168, 1, 30 };
Server server(80);

int numPins = 5;
int pins[] = {3, 4, 5, 6, 7};
```

```
int pinState[] = {0, 0, 0, 0, 0};
char line1[100];

void setup()
{
  for (int i = 0; i < numPins; i++)
  {
    pinMode(pins[i], OUTPUT);
  }
  Serial.begin(9600);
  Ethernet.begin(mac, ip);
  server.begin();
}

void loop()
{
  Client client = server.available();
  if (client)
  {
    while (client.connected())
    {
      readHeader(client);
      if (! pageNameIs("/"))
      {
        client.stop();
        return;
      }
      client.println("HTTP/1.1 200 OK");
      client.println("Content-Type: text/html");
      client.println();

      // send the body
      client.println("<html><body>");
      client.println("<h1>Output Pins</h1>");
      client.println("<form method='GET'>");
      setValuesFromParams();
      setPinStates();
      for (int i = 0; i < numPins; i++)
      {
        writeHTMLforPin(client, i);
      }
      client.println("<input type='submit' value='Update'/>");
      client.println("</form>");
      client.println("</body></html>");

      client.stop();
    }
  }
}

void writeHTMLforPin(Client client, int i)
{
```

```
  client.print("<p>Pin ");
  client.print(pins[i]);
  client.print("<select name='");
  client.print(i);
  client.println("'>");
  client.print("<option value='0'");
  if (pinState[i] == 0)
  {
    client.print(" selected");
  }
  client.println(">Off</option>");
  client.print("<option value='1'");
  if (pinState[i] == 1)
  {
    client.print(" selected");
  }
  client.println(">On</option>");
  client.println("</select></p>");
}

void setPinStates()
{
  for (int i = 0; i < numPins; i++)
  {
    digitalWrite(pins[i], pinState[i]);
  }
}

void setValuesFromParams()
{
  for (int i = 0; i < numPins; i++)
  {
    pinState[i] = valueOfParam(i + '0');
  }
}

void readHeader(Client client)
{
  // read first line of header
  char ch;
  int i = 0;
  while (ch != '\n')
  {
    if (client.available())
    {
      ch = client.read();
      line1[i] = ch;
      i ++;
    }
  }
  line1[i] = '\0';
  Serial.println(line1);
}
```

```
boolean pageNameIs(char* name)
{
  // page name starts at char pos 4
  // ends with space
  int i = 4;
  char ch = line1[i];
  while (ch != ' ' && ch != '\n' && ch != '?')
  {
    if (name[i-4] != line1[i])
    {
      return false;
    }
    i++;
    ch = line1[i];
  }
  return true;
}

int valueOfParam(char param)
{
  for (int i = 0; i < strlen(line1); i++)
  {
    if (line1[i] == param && line1[i+1] == '=')
    {
      return (line1[i+2] - '0');
    }
  }
  return 0;
}
```

The sketch uses two arrays to control the pins. The first, **pins**, just specifies which pins are to be used. The **pinState** array holds the state of each pin: either 0 or 1.

To get the information coming from the browser form about which pins should be on and which should be off, it is necessary to read the header coming from the browser. In fact, all you need is contained in the first line of the header. You will use a character array **line1** to contain the first line of the header.

When the user clicks on the Update button and submits the form from the browser, the URL for the page will look something like this:

```
http://192.168.1.30/?0=1&1=1&2=0&3=0&4=0
```

The request parameters come after the **?** and are each separated by an **&**. Looking at the first parameter (**0=1**), this means that the first pin in the array (**pins[0]**) should have the value 1. If you were to look at the first line of the header, you would see those same request parameters there:

```
GET /?0=1&1=1&2=0&3=0&4=0 HTTP/1.1
```

Before the parameters, there is the text **GET /**. This specifies the page requested by the browser. In this case, / indicates the root page.

In the loop of the sketch, you call the **readHeader** function to read the first line of the header. You then use the **pageNameIs** function to check that the page request is for the root page /.

The sketch then generates the header and the start of the HTML form that is to be displayed. Before writing the HTML for each of the pins, the sketch calls the **setValuesFromParams** function to read each of the request parameters and set the appropriate values in the **pinStates** array. This array is then used to set the values of the pin outputs before the **writeHTMLforPin** function is called for each of the pins. This function generates a selection list for each pin. It has to build this list part by part. The **if** statements ensure that the appropriate options are selected.

The functions **readHeader**, **pageNameIs**, and **valueOfParam** are useful general-purpose functions that you can make use of in your own sketches.

You can use your multimeter as you did in Chapter 6 to verify that the pins are indeed turning on and off. If you are feeling more adventurous, you could attach LEDs or relays to the pins to control things.

Conclusion

Having used shields and associated libraries in the last two chapters, it is now time to investigate the features that enable libraries to be written and learn how to write libraries of your own.

11

C++ and Libraries

Arduinos are simple microcontrollers. Most of the time, Arduino sketches are quite small, so using the C programming language works just fine. However, the programming language for Arduino is actually C++ rather than C. C++ is an extension to the C programming language that adds something called *object orientation*.

Object Orientation

This is only a short book, so an in-depth explanation of the C++ programming language is beyond its scope. The book can, however, cover the basics of C++ and object orientation. But the main goal is to increase the *encapsulation* of your programs. Encapsulation keeps relevant things together, something that makes C++ very suitable for writing libraries such as those that you have used for the Ethernet and LCD sketches in earlier chapters.

There are many good books on the topics of C++ and object-oriented programming. Look for the higher-rated books on the topic in your favorite online bookstore.

Classes and Methods

Object orientation uses a concept called *classes* to aid encapsulation. Generally, a class is like a section of a program that includes both variables—called *member variables*—and *methods*, which are like functions but apply to the class. These functions can either be *public*, in which case the methods and functions may be used by other classes, or *private*, in which case the methods can be called only by other methods within the same class.

Whereas an Arduino sketch is contained in a single file, when you are working in C++, you tend to use more than one file. In fact, there are generally two files for every class: A *header file*, which has the extension .h, and the *implementation file*, which has the extension .cpp.

Built-in Library Example

The LCD library has been used in the two previous chapters, so let's look more closely and see what is going on in a little more detail.

Referring back to sketch 9-01 (open this in your Arduino IDE), you can see that the **include** command includes the file LiquidCrystal.h:

```
#include <LiquidCrystal.h>
```

This file is the header file for the class called **LiquidCrystal**. This file tells the Arduino sketch what it needs to know to be able to use the library. You can actually retrieve this file if you go to your Arduino installation folder and file and find the file libraries/LiquidCrystal. You will need to open the file in a text editor. If you are using a Mac, then right-click on the Arduino app itself and select the menu option Show Package Contents. Then navigate to Contents/Resources/Java/libraries/LiquidCrystal.

The file LiquidCrystal.h contains lots of code, as this is a fairly large library class. The code for the actual class itself, where the nuts and bolts of displaying a message actually reside, are in the file LiquidCrystal.cpp.

In the next section, a simple example library will be created that should put the concepts behind a library into context.

Writing Libraries

Creating an Arduino library might seem like the kind of thing that only a seasoned Arduino veteran should attempt, but actually it is pretty straight-forward to make a library. For example, you can convert into a library the

flash function from Chapter 4 that causes an LED to flash for a specified number of times.

To create the C++ files that are needed to do this, you will need a text editor for your computer—something like TextPad on Windows or Text-Mate on Mac.

The Header File

Start by creating a folder to contain all the library files. You should create this folder inside the libraries folder of your Arduino documents folder. In Windows, your libraries folder will be in My Documents\Arduino. On the Mac, you will find it in your home directory, Documents/Arduino/, and on Linux, it will be in the sketchbook directory of your home directory. If there is no libraries folder in your Arduino, then create one.

This libraries folder is where any libraries you write yourself, or any "unofficial" contributed libraries, must be installed.

Call the folder Flasher. Start the text editor and type the following into it:

```
// LED Flashing library

#include "WProgram.h"

class Flasher
{
  public:
    Flasher(int pin, int duration);
    void flash(int times);
  private:
    int _pin;
    int _d;
};
```

Save this file in the Flasher folder with the name Flasher.h. This is the header file for the library class. This file specifies the different parts of the class. As you can see, it is divided into public and private parts.

The public part contains what looks like the start of two functions. These are called methods and differ from functions only insofar as they are associated with a class. They can be used only as part of the class. Unlike functions, they cannot be used on their own.

The first method, **Flasher**, begins with an uppercase letter, which is something you would not use with a function name. It also has the same name as the class. This method is called a *constructor*, which you can apply to create a new **Flasher** object to use in a sketch.

For example, you could put the following in a sketch:

```
Flasher slowFlasher(13, 500);
```

This would create a new **Flasher** called **slowFlasher** that would flash on pin D13 with a duration of 500 milliseconds.

The second method in the class is called **flash**. This method takes a single argument of the number of times to flash. Because it is associated with a class, when you want to call it, you have to refer to the object that you created earlier, as follows:

```
slowFlasher.flash(10);
```

This would cause the LED to flash ten times at the period that you specified in the constructor to the **Flasher** object.

The private section of the class contains two variable definitions: one for the pin, and one for the duration, which is simply called d. Every time that you create an object of class **Flasher**, it will have these two variables. This enables it to remember the pin and duration when a new **Flasher** object is created.

These variables are called member variables because they are members of the class. Their names generally are unusual in that they start with an underscore character; however, this is just a common convention, not a programming necessity. Another commonly used naming convention is to use a lowercase m (for *member*) as the first letter of the variable name.

The Implementation File

The header file has just defined what the class looks like. You now need a separate file that actually does the work. This is called the implementation file and has the extension .cpp.

So, create a new file containing the following and save it as Flasher.cpp in the Flasher folder:

```
#include "WProgram.h"
#include "Flasher.h"

Flasher::Flasher(int pin, int duration)
{
  pinMode(pin, OUTPUT);
  _pin = pin;
  _d = duration / 2;
}

void Flasher::flash(int times)
{
  for (int i = 0; i < times; i++)
  {
    digitalWrite(_pin, HIGH);
    delay(_d);
    digitalWrite(_pin, LOW);
    delay(_d);
  }
}
```

There is some unfamiliar syntax in this file. The method names are both prefixed by **Flasher::**. This indicates that the methods belong to the **Flasher** class.

The constructor method (**Flasher**) just assigns each of its parameters to the appropriate private member variable. The **duration** parameter is divided by two before being assigned to the member variable _d. This is because the delay is called twice, and it seems more logical for the duration to be the total duration of the flash and the gap between flashes.

The **flash** function actually carries out the business of flashing; it loops for the appropriate number of times, turning the LED on and off for the appropriate delay.

Completing Your Library

You have now seen all of the essentials for completing the library. You could now deploy this library and it would work just fine. However, there are two further steps that you should take to complete your library. One is to define the keywords used in the library so that the Arduino IDE can show them in the appropriate color when users are editing code. The other is to include some examples of how to use the library.

Keywords

To define the keywords, you have to create a file called keywords.txt, which goes into the Flasher directory. This file contains just the two following lines:

```
Flasher    KEYWORD1
flash      KEYWORD2
```

This is essentially a two-column table in a text file. The left column is the keyword and the right column an indication of the type of keyword it is. Class names should be a **KEYWORD1** and methods should be **KEYWORD2**. It does not matter how many spaces or tabs you put between the columns, but each keyword should start on a new line.

Examples

The other thing that you, as a good Arduino citizen, should include as part of the library is a folder of examples. In this case, the library is so simple that a single example will suffice.

The examples must all be placed in a folder called examples inside the Flasher folder. The example is in fact just an Arduino sketch, so you can create the example using the Arduino IDE. But first, you have to quit and then reopen the Arduino IDE to make it aware of the new library.

After restarting the Arduino IDE, from the Arduino IDE's menu, select File and then New to create a new sketch window. Then from the Menu, select Sketch and the Import Library option. The Options should look something like Figure 11-1.

The libraries above the line in the submenu are the official libraries; below this line are the "unofficial" contributed libraries. If all has gone well, you should see Flasher in the list.

If Flasher is not in the list, it is very likely that the Flasher folder is not in the libraries folder of your sketches folder, so go back and check.

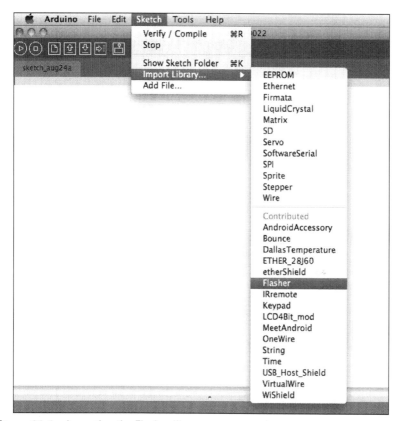

Figure 11-1   *Importing the Flasher library*

Type the flowing into the sketch window that has just been created:

```
#include <Flasher.h>

int ledPin = 13;
int slowDuration = 300;
int fastDuration = 100;

Flasher slowFlasher(ledPin, slowDuration);
Flasher fastFlasher(ledPin, fastDuration);

void setup(){}

void loop()
```

```
{
    slowFlasher.flash(5);
    delay(1000);
    fastFlasher.flash(10);
    delay(2000);
}
```

The Arduino IDE will not allow you to save the example sketch directly into the libraries folder, so save it somewhere else under the name Simple Flasher Example and then move the whole Simple Flasher Example folder that you just saved into the examples folder in your library.

If you restart your Arduino IDE, you should now see that you are able to open the example sketch from the menu as shown in Figure 11-2.

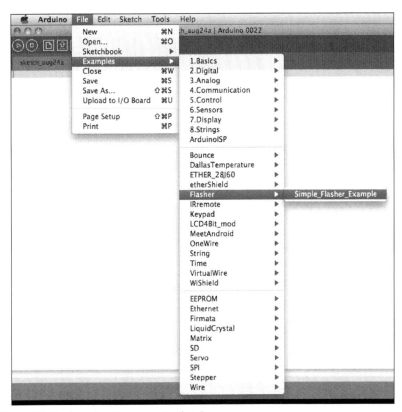

Figure 11-2 *Opening the example sketch*

Conclusion

There is more to C++ and to writing libraries, but this chapter should get you started. It should also be sufficient for most of what you are likely to do with an Arduino. These Arduinos are small devices and the temptation is often to overengineer solutions that could otherwise be very simple and straightforward.

That concludes the main body of this book. For further information on Arduino and where to go next, a good starting point is always the official Arduino website at www.arduino.cc. Also, please refer to the book's website at www.arduinobook.com, where you will find errata and other useful resources.

If you are looking for help or advice, the Arduino community on www.arduino.com/forum is extremely helpful. You will also find the author on there with the username Si.

Index

Symbols

>= (greater than or equal to) comparison operator, 44

<= (less than or equal to) comparison operator, 44

– (minus) operator, 42

!= (not equal to) comparison operator, 44

/ (slash), as division operators, 42

; (semicolon), in programming syntax, 29

| | (or) operator, 60

+ (plus), as addition operator, 42

= (assignment) operator, assigning values to variables, 40

== (equal to) comparison operator, 44, 60

&& (and) operator, manipulating values, 60

* (asterisk), as multiplication operator, 42

[] (square brackets), in array syntax, 68

<< (bit shift operator), 119

< (less than) comparison operator, 44

> (greater than) comparison operator, 44

A

abs function, math functions in library, 108

addition (+) operator, 42

Algorithms + Data Structures = Programs (Wirth), 67

alphanumeric LCD Shield, 126

analog inputs, 4–5, 102–103

analog outputs, 100–102

analogRead function, 102

analogWrite function, 102

and (&&) operator, manipulating values, 60

Arduino Bluetooth, 11–12

Arduino Diecimila, 9

Arduino Duemilanove, 9

Arduino Lilypad, 11–12

Arduino Mega, 10

Arduino Nano, 10–11

Arduino, origins of, 7

Arduino Uno

    in Arduino family of development boards, 9

    ATMega328 processor in, 117

arguments

    passing to functions, 28–29

    of random function, 106

    of tone function, 111

arithmetic

    numeric variables and, 40–42

    operators, 42

arrays

    overview of, 67–71

    PROGMEM and, 116–117

    SOS signal example, 71–72

    string arrays. *See* strings

    for translating Morse code. *See* Morse code translator

ASCII code, 73

assignment (=) operator, assigning values to variables, 40

asterisk (*), as multiplication operator, 42

ATmega1280, Mega and, 10

ATmega168, Uno and, 9

153

31436803R00100

Made in the USA
San Bernardino, CA
09 March 2016